Climate Change Mitigation and Adaptation—ZEMCH 2016

Special Issue Editor
Arman Hashemi

MDPI • Basel • Beijing • Wuhan • Barcelona • Belgrade

MDPI

Special Issue Editor
Arman Hashemi
University of Brighton
UK

Editorial Office
MDPI
St. Alban-Anlage 66
Basel, Switzerland

This edition is a reprint of the Special Issue published online in the open access journal *Sustainability* (ISSN 2071-1050) in 2017 (available at: http://www.mdpi.com/journal/sustainability/special_issues/ ZEMCH_2016).

For citation purposes, cite each article independently as indicated on the article page online and as indicated below:

Lastname, F.M.; Lastname, F.M. Article title. *Journal Name* **Year**, *Article number*, page range.

First Editon 2018

Cover photo courtesy of Dr. Arman Hashemi.

ISBN 978-3-03842-965-4 (Pbk)
ISBN 978-3-03842-966-1 (PDF)

Table of Contents

About the Special Issue Editor

Arman Hashemi, Senior Lecturer in Architectural Technology, has several years of experience in practice and academia in the UK and overseas. He has been involved in numerous research, design and construction projects and has worked on a range of award-winning architectural projects. Dr Hashemi's primary research is on building performance evaluation with a focus on energy efficiency, indoor air quality and thermal comfort in buildings. His research interests also include offsite/modern methods of construction and sustainable low-income housing. Dr. Hashemi's research and development projects have been widely recognized and commended for their potential impact on the construction industry. He has received two CIOB I&R awards for his research quality and his internationally patented product, the Advanced Thermal Shutter System. The results of his research have been documented in numerous publications including refereed journal articles, conference papers, edited books, book chapters and patents. Dr. Hashemi is a Fellow of the Higher Education Academy of the UK, a Member of the Chartered Management Institute, and a Member of the Iranian Construction Engineers Organization. He has also acted as scientific chair, guest editor and as a member of editorial boards of scientific committees of several international journals and conferences.

Preface to "Climate Change Mitigation and Adaptation—ZEMCH 2016"

The adverse effects of climate change are becoming more evident and global warming is now undeniable. Efforts to mitigate the impacts of climate change have been progressing slowly and adaptation is now one of the major strategies being considered by both developed and developing countries. The challenging and ambitious targets of the Paris Agreement in December 2015 were a milestone in such efforts; however, despite the commitment of many countries, greenhouse gas emissions have not been decreasing sufficiently to slow down the global warming. Indeed, there are currently debates on whether reducing the emissions to 40 gigatonnes by 2030 and limiting the temperature rise to "1.5 ○C above pre-industrial levels" are still realistic and achievable. Nevertheless, reducing CO_2 emissions from the construction industry would significantly contribute to achieving the objectives of the Paris Agreement.

Passive design, low/zero energy buildings and energy efficient refurbishment of existing buildings are some of the strategies which can effectively reduce CO_2 emissions in the construction industry. Yet, the excessive attention to energy efficiency together with poor design/construction and short-term governmental policies in response to global and social pressures to meet ambitious targets, without much attention to long-term consequences, have led to some serious issues such as poor indoor air quality, chronic overheating and other associate problems, affecting the health and wellbeing of the occupants of energy efficient buildings.

For various reasons, attention is gradually shifting towards developing countries that are becoming major contributors to greenhouse gas emissions. Unlike developed countries that have the resources to invest and respond to climate change, adaptation in poor countries is left to the individuals to mitigate the risks of climate change. Low-income populations living in the least developed countries are hit the worst by climate change due to their vulnerable living conditions and lack of access to an appropriate infrastructure as well as low level of awareness and knowledge to deal with the consequences of climate change and global warming.

To this end, this book explores the current issues around sustainable design and energy efficiency within the construction industry, in both developed and developing countries, aiming to contribute to global efforts towards climate change mitigation and adaptation.

Arman Hashemi
Special Issue Editor

sustainability

MDPI

Review

A Review on Building Integrated Photovoltaic Façade Customization Potentials

Daniel Efurosibina Attoye *, Kheira Anissa Tabet Aoul and Ahmed Hassan

Department of Architectural Engineering, United Arab Emirates University, Al Ain 15551, UAE;
kheira.anissa@uaeu.ac.ae (K.A.T.A.); ahmed.hassan@uaeu.ac.ae (A.H.)
* Correspondence: 201590088@uaeu.ac.ae; Tel.: +971-52-971-6472

Received: 15 September 2017; Accepted: 22 November 2017; Published: 9 December 2017

Abstract: Technological advancement in Building Integrated Photovoltaics (BIPV) has converted the building façade into a renewable energy-based generator. The BIPV façade is designed to provide energy generation along with conventional design objectives such as aesthetics and environmental control. The challenge however, is that architectural design objectives sometimes conflict with energy performance, such as the provision of view and daylight versus maximum power output. In innovative cases, the characteristics of conventional BIPV façades have been modified by researchers to address such conflicts through customization as an emerging trend in BIPV façade design. Although extensive reviews exist on BIPV product types, design integration, adoption barriers and performance issues, research on BIPV customization has not been reviewed as a solution to BIPV adoption. This paper seeks to review the potential of BIPV façade customization as a means of enhancing BIPV adoption. The current paper identifies customization parameters ranging from the customization category, level, and strategies, and related architectural potential along with an assessment of their impact. The findings reflect that elemental and compositional level customization using combined customization strategies provide enhanced BIPV products. These products are well integrated for both energy generation and aesthetic applications with a power output increase of up to 80% in some cases. The paper concludes that a wide range of BIPV adoption barriers such as aesthetics, architectural integration, and performance can be overcome by appropriate BIPV customization.

Keywords: Building Integrated Photovoltaics (BIPV); façade; customization; architectural potentials; barriers

---

1. Introduction

Buildings are a main source of global energy consumption and CO_2 emissions; accounting for about 40% of global energy consumption [1,2] The international contribution to sustainability has generated a large number of publications in relevant journals and conferences over the last four decades [3] and has established a dire need to reduce greenhouse gas emissions [4–6] as these gases are potential causes of threats to the ecosystem such as global warming [7–10]. From the Kyoto protocol of 1997 to the Paris Agreement of 2015, various policy directions have been motivated to mitigate international environmental pollution. Mitigation in various dimensions is a key factor to improving the environment for future generations [11,12]. At the building scale, the potential of on-site renewable energy generation to optimize energy demand and supply infrastructure has been investigated [13–15]. This provides an opportunity to address environmental pollution; which has frequently been linked with the rising level of nonrenewable energy consumption [5]. Building Integrated Photovoltaics (BIPV) provides such an opportunity through clean micro-energy generation being adoptable to various building designs. Several studies indicate that application of BIPV leads to substantial energy savings [16–18] and thus related gains in energy consumption and reduction of pollution sources.

BIPV reduces the damage done to the ecosystem through conventional energy sources [19] and is a promising way of relieving the increasing financial and environmental costs of fossil fuel energy generation [20]. Technological advancements have evolved BIPV into a PV application with the capability of electrical delivery at a comparatively lower cost than grid electricity for certain end users in certain peak demand niche markets [21]. As a contemporary material available to architects, BIPV serves simultaneously as a part of the building envelope and an energy source. BIPV systems can be more cost effective simply because their composition and location replaces a number of conventional components and thus provide multiple gains which are reviewed in details in this paper. These include savings in materials and electricity costs, reduced use of fossil fuels, decreasing carbon and greenhouse gases emissions and improved architectural image of the building [22,23].

However, several studies highlight various barriers which are limitations to the widespread adoption of BIPV [24–34]. They range from general product issues such as performance, aesthetics and technical complexity [26] to specific regional issues such as the need for extensive education on professional and public levels [24,25,27]. Greater attention to research and development in customization and BIPV product designs with good architectural aesthetics and integrality have been suggested by reviewers as potential solutions to these [25,26,28,32]. However, our survey of recent BIPV reviews over the last 5 years (Table 1) shows that there is only limited information on customization as potential driver for BIPV adoption. Only partial attention has been made of custom BIPV; relating to mention of strategies [35,36] and cost limitations [37]. In two other cases [38,39] a descriptive inventory of several market-ready custom BIPV product applications connecting cell technology and architectural integration is given [38]. Also, details on the possibilities, market options, and aesthetic levels of customizability were presented [39].

Table 1. Summary of customization content in recent Building Integrated Photovoltaics (BIPV) reviews.

Reference	Title/Focus	Customization-Related Content
[40]	Recent advancement in BIPV product technologies: a review	-
[41]	Embedding passive intelligence into building envelopes: a review	Reference to a system-based process design
[35]	A critical review on building integrated photovoltaic products and their applications	Brief mention
[22]	Double skin façades (DSF) and BIPV: a review of configurations and heat transfer characteristics	Inference to different design modes
[42]	A comprehensive review on design of building integrated photovoltaic system	Reference made to an energy-conscious process design
[39]	Overview and analysis of current BIPV products: new criteria for supporting the technological transfer in the building sector	Possibilities, market options, aesthetic levels; an architectural layering process design approach
[43]	PV glazing technologies	-
[36]	Building Integrated Photovoltaics: a Concise. Description of the Current State of the Art and Possible Research Pathways	Brief mention
[37]	Building Integrated Photovoltaics (BIPV): review, Potentials, Barriers and Myths	Brief mention of the need, possibilities and challenges
[38]	'State-of-the-art' of building integrated photovoltaic products	Details on available custom products in the market
[44]	Building integrated photovoltaic products: a state-of-the-art review and future research opportunities	Possibilities and available custom products in the market
[45]	The path to the building integrated photovoltaics of tomorrow	Brief mention of possible future in product variety
[46]	Whole systems appraisal of a UK Building Integrated Photovoltaic (BIPV) system: energy, environmental, and economic evaluations	-
[47]	Photovoltaics and zero energy buildings: a new opportunity and challenge for design	-
[48]	Architectural Quality and Photovoltaic Products	Mention of examples, function and challenges

Source: By Authors.

Furthermore, no information or review of experimental investigations on customization is presented to theorize its description or justify its potentiality. This review paper aims to fill this gap as identified in the literature by investigating the characteristics, strategies and potentials of BIPV customization. In addition, the study seeks to showcase the opportunities provided by customization to address the barriers of conventional BIPV.

1.1. BIPV Customisation: Working Definition

Customization is, "the action of modifying something to suit a particular individual or task" [49]. It can also be described as the configuration of products and services to meet customers' individual needs [50]. These definitions suggest that customization is directly associated with the identification of a function, need or objective. As it relates to BIPV, several customization objectives have been investigated, such as aesthetics [51,52], architectural integration [53,54], thermal management [55–57], and shading [15,41,53]. These objectives consequently determine the added function of the designed custom BIPV façade along with energy production from the solar cells. Ref. [39,58] suggest that BIPV designs can follow a systematic design process. This infers that various levels/stages of customization are identifiable; [39] suggests a cell, module, and façade level activity while [58] presents an elemental, compositional, and integrational level of interest. In both representative cases, the idea is to first customize the cell, then the module, and finally the façade.

1.2. Research Design

The present review is divided into three main sections; first an overview on BIPV façades, then an appraisal of standard BIPV barriers, and finally a review of BIPV façade customization studies. Data collection and analysis steps of relevant studies for all sections of this paper were limited to English-language studies found in the ScienceDirect and Google Scholar database. In Section 1, an assessment of the mention given to BIPV customization in previous state-of-the art reviews was presented to validate the need for this investigation. For Section 2, we identified eleven (11) studies within the past five years which focused primarily on a review of barriers inherent to BIPV in general, BIPV products or to BIPV adoption. The selection was limited to the last five years, as BIPV is an evolving technology and this review seeks to identify current mitigating issues. These studies collectively represent the views of several researchers drawn from surveys involving close to 1000 respondents, based on experiences and findings from professionals and researchers worldwide. In Section 3, keywords such as "BIPV customization", "custom BIPV", "customized BIPV" were used in our search, but at the time of writing this review, no studies with these exact words were found. We expanded our search for related titles on BIPV façade customization and identified 25 representative studies with related abstracts and thus focused our investigation on these. Figure 1 shows a color-coded mind map for this investigation; it reflects the research direction and connections, as well as a basis for deductive reasoning which informs the resulting conclusions made in this review. The blue-coded section groups together the research on BIPV types and potential benefits which are discussed in detail later in Section 2 (Overview on BIPV façades). The red coded section combines the barriers that affect BIPV adoption into the built environment later discussed separately in Section 3 (BIPV barriers). The green coded sections put together the specifics of customization as an approach to enhance BIPV adoption into built environment later discussed in Sections 4.1 and 4.2 (BIPV customization investigations). The detailed discussions on the specific findings of the mind map are discussed in the following Sections 2–4.

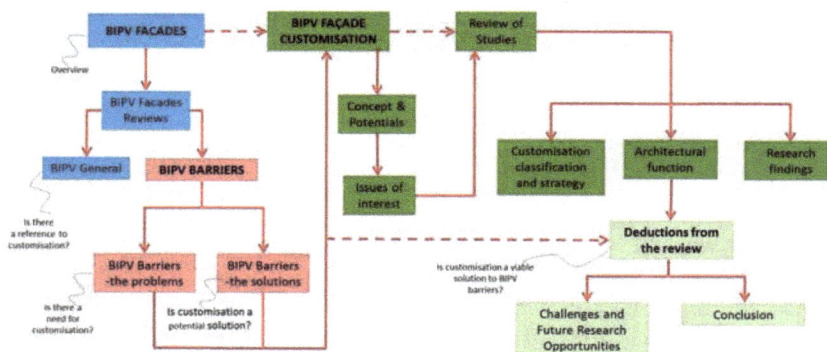

Figure 1. BIPV façade customization review mind map. Source: authors.

2. Overview on BIPV Façade Applications

The building façade is conventionally made up of walls, glazing, cladding and fenestrations; and other structures like shading devices, parapets and balconies. Each of these building components provide opportunities for integrating PVs to the building and by extension, for façade customization [36,37,59–62]. The main BIPV façade applications extracted from literature [39,43,59,61,63] include curtain walls, glazing, external/shading devices, and innovative applications. Table 2 presents an overview of these applications using representative built examples to describe the advantages and disadvantages of each of these types.

Table 2. Design impact of BIPV façade types. Source: authors.

BIPV Façade Type	Design Impact
1. Curtain Wall/Cladding Systems Solar panels integrated as a conventional cladding system for curtain walls and single layer façades [37]. a.	• Advantages - Intelligent way of balancing daylighting and shading [37]. - Iconic importance in the field of architecture [37]. - Different colors and visual effects can be included [61]. - Regulates the internal temperatures of the building by minimizing solar gain in the summer [61]. - Light effects from these panels lead to an ever-changing pattern of shades in the building itself [61]. - Impacts on overall architectural image - Maximizes façade wall for energy generation • Disadvantages - Installation costs can be high [61]. - Potentially less energy than on roof-top [37]. - Requires complex planning and compliance with a great many physical properties [37]. - Properly handling needed to prevent view obstruction by electrical cables
2. Solar Glazing and Windows Applied as semi-transparent/translucent parts of the façade based on solar cell transparency. They can be integrated into windows, glazing panels, for view or daylighting [59]. b.	• Advantages - Allowing for filtered view as well as energy generation [61]. - Potential application as opaque or semi-transparent/translucent glazing [59] - Special PV elements used thermal insulators in combination with standard double or triple glazing elements [59]. - Added functionality as sun shading - The patterns from the shading generate a dynamic experience of spatial variety through the day. • Disadvantages - Potentially lower efficiencies [37] - Increase in cell spacing yields less energy due to fewer cells

Table 2. *Cont.*

BIPV Façade Type	Design Impact
3. External Devices/Accessories	• Advantages

Sunshades and sunscreens, spandrels, balconies parapets, elements of visual and acoustic shielding [61].	- Potential for minimizing both building heat loads and energy consumption [59].
	     - Vertical or horizontal sun shading provided above windows [61]
	     - Use of building shading structure as mounting to prevent additional load on façade [61].
	     - Potential as fixed or adjustable devices [59,61]
	     - Allows for PV modules of different shapes [61].

 c.
	• Disadvantages
	     - Shadows cast from BIPV panels may need filtering to even out light distribution
	     - Obstruction of view if not transparent or translucent

4. Advanced/Innovative Envelope Systems	• Advantages

Such as double skin façades, active skins, rotating or moving façade parts, etc. [59]	- Integration with advanced aesthetic polymer technologies [59].
	     - Generation of heat in winter for space heating
	     - Double skin façades assist in cooling of BIPV panels [64].
	     - Possible integration with other building elements for performance and aesthetics [53,55–57]

 d.
	• Disadvantages
	     - Potentially more expensive than other types
	     - Energy maybe required for extraction of heat in summer via mechanical means or forced ventilation [64].

a. Curtain Wall of Hanergy Office, Guangdong, China; showing BIPV cladding; b. KTH Executive School AB, Sweden; showing glazing with wide spaced solar cells for daylighting and view; c. Shading devices on Kingsgate House London, UK; showing vertical polycrystalline panels; d. Innovative façade of Hanergy Headquarters, Beijing, China; showing innovative "dragon scale" arrangement of BIPV modules

2.1. Strategic Benefits of BIPV

BIPV is a multifunctional technology and they are therefore usually designed to serve more than one function [36,60,65]. Along with the fundamental function of producing of electricity, the multi-functionality of BIPV thus implies that it can fulfill several other tasks as a façade element such as solar protection and glare protection. Three identified classes of such added function or benefits from literature relate to the building envelope design, economic advantages and environmental impact.

- Design Benefits: relating to architectural integration and function of BIPV as a building component
- Economic Benefits: relating to financial advantages accrued as a result of BIPV application
- Environmental Benefits: relating to micro or macro environment improvements due to BIPV application

The list below contains a categorization of the multiple functions that BIPV modules can perform based on its unique characteristics.

1. Design-related benefits

    a. View and daylighting—semi-transparent options allow for light transmission and contact with exterior [61,66,67].
    b. Aesthetic quality—integration in buildings as a design element [36,61]
    c. Sun protection/shadowing/shading modulation—used as fixed or tracking shading devices [36,37,60,61,67],
    d. Replacement of conventional materials such as brickwork [37].
    e. Public demonstration of owner's green ecological and future-oriented image [61].

    f.   Safety—applied as safety glass [61].

    g.   Noise protection—reaching up to 25 dB sound dumping [36,37,61,67].

    h.   Heat protection/Thermal insulation (heating as well as cooling)—improving the efficiency of cells by cooling through rear ventilation [36,37,61]

    i.   Visual cover/refraction—one-way mirroring visual cover [60,61].

2.   Economic Benefits

    a.   Removal of the need for the transmittance of electricity over long distances from power generation stations [68,69].

    b.   Reduction in capital expenditure for infrastructure and maintenance [68,69].

    c.   Reduction in land use for the generation of electricity [28,70].

    d.   Material and labour savings as well as electrical cost reductions [36].

    e.   Reduction in additional assembly and mounting costs, leading to on-site electricity and lowering of total building material costs and significant savings [45]. In addition, ongoing costs of a building are reduced via operational cost savings and reduced embodied energy [71].

    f.   Combined with grid connection, FITs; cost savings equivalent to the rate the electricity is close to zero [28,46,72]

3.   Environmental Benefits

    a.   Reduction of carbon emissions [28]

    b.   The pollution-free benefit of solar energy [45].

    c.   Reduces the Social Cost of Carbon (SCC) relating to the health of the public and the environment [28].

3. BIPV Façade Applications: Barriers and Strategies

Notwithstanding the stated multi-functionality of BIPV already expressed in several studies, its adoption is globally challenged by certain barriers. It has been argued that sustainability goals of the future can only be achieved if we look beyond new technologies themselves, and account for the complex human factors influencing their adoption and use [73]. Several researchers have investigated these barriers and their studies show that there are various perspectives and issues of concern. These include challenges in the various stages of application [30] such as the design stage and installation stage, and in some regional cases, expertise limitation, lack of promotion, and financial issues [27]. There are also key barriers that are general to BIPV adoption, and in some cases, affect the building integration of other renewable energy technologies. Some of these general issues from a more holistic point of view are sociotechnical, management, economic, and policy-related [29] as well as knowledge and information-related [26]. Others include insufficient presentation of BIPV product and project databases, lack of adequate business models, and insufficient dissemination of BIPV information [24].

In almost all of these studies, strategies for overcoming these barriers have been proposed. These strategies are drivers in various forms with the potential to advance or facilitate the BIPV implementation in the built environment. In some cases, they are proposed solutions to counter one or more barriers when fully applied. Table 3 gives a detailed overview of the findings of these related studies showing a categorization of the barriers and drivers identified to BIPV application; to clearly identify the issues of concern and potential solutions.

Table 3. A detailed overview of identified BIPV barriers and drivers. Source: By Authors.

Barriers	Drivers
1. Product efficiency and design	1. Research & Development on product design and design tools
• System performance [25–27,29–31,34] • Design standards, codes and regulations [24,28,30,32,34] • Design tools and software [25,26,30,33] • Aesthetics and architectural integration [25,26,30–33]	• Enhanced product design for architectural integration, innovative manufacturing, customization, standardization and modularity [24,26,28,30,31,33] • Improved product performance [25,26,28,31] • Development and application of Building Information Modelling (BIM), simulation & mathematical software and tools for design, performance monitoring and environmental issues [24,26,28,30,31,33] • International research and design collaborations [24,26,28]
2. Product and project demonstration and databases [24–26]	2. Educational programs and public awareness projects
3. Education	• Professional technical experience, training and development [28–30,32,34] • Development of an international Product and Project database [24,26] • Development of educational material for universities [25,33]. • Increased public knowledge via effective marketing; outreach events; use of specific communication tools for client motivation [24–31,34] • Urban demonstration projects on BIPV and energy-related issues [28,32,34]
• Professional training and expertise [24–27,29,30,32–34] • Public awareness and perception [24–29,32–34]	
4. Economy	3. Active governmental interventions
• Material and system costs [25–29,32–34] • Governmental support and policies [26–29,31,33] • International or bank support [27,29,31]	• Dedicated government support and incentives [27,29–34] • BIPV implementation policy formation; [28,29,31] • Non-financial incentives as 'green accreditation' and reduction in lending rates [27,28]
5. Gap between PV and building industry [29,30,32]	4. International professional management and collaborations
6. Management & business and project planning [24,28,29,32]	• Increased collaboration between government, research bodies, manufacturers, building professionals and clients [26,28–32,34] • Development of specific management and business models [24,26,30] • Development of international guidelines, standards and codes for BIPV implementation [24,26,30,31]

The collective information from these 11 studies represents the opinions of close to 1000 international respondents. The summary of these findings was distilled and diagrammatically presented in a force field analysis (Figure 2) for further scrutiny. A force field analysis is a management analytical tool used to conceptualize the forces interacting to promote and oppose change in a given situation [34]. We have applied it to give a visual representation of the barriers, stated as restraining factors and drivers as facilitating factors. Kurt Lewin is often acknowledged as the first to propose this technique in 1951 [74]. The weight of the arrows in this adaptation is shown in percentages, and obtained from the frequency of mention in studies of a barrier or solution.

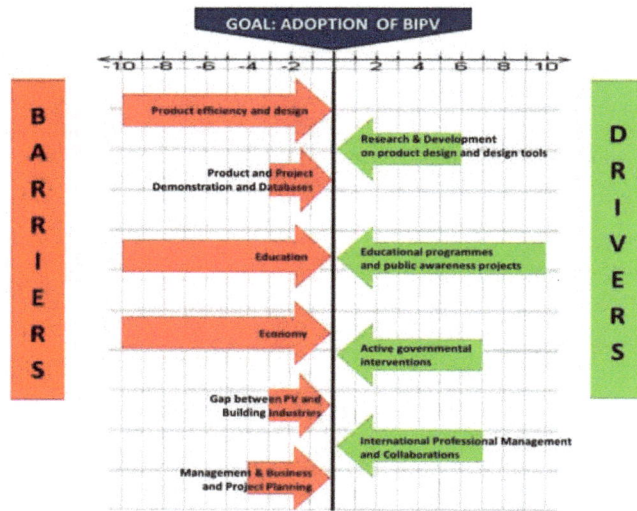

Figure 2. Simplified force field analysis of barriers and drivers of BIPV adoption. Source: By Authors.

3.1. Force Field Analysis: Comparison of Barriers and Strategies

Statistically speaking, six classes of barriers relating to the product, education, economy, and industry were identified in the referred literature with further sub-division of three of these classes. As an example, the product efficiency and design class encompasses related barriers such as system performance, design standards, codes and regulations. It also includes design tools and software, aesthetics and architectural integration issues. As observed from our investigation, the need to address public awareness and perception [24–29,32–34] and the insufficient professional training and expertise [24–27,29,30,32–34] are the most frequently identified barriers to BIPV adoption. This suggests an international agreement that the need for proper education regarding the potentials of BIPV is lacking in both public and professional domains. Comparatively, insufficient product and project demonstration and database, as well as insufficient international or bank support are ranked as the least identified barriers. It may be assumed therefore that client motivation via these latter support schemes may not be directly related to the reluctance to BIPV adoption. Another deduction from this survey is that comparatively, there are potentially more product efficiency and design related barriers, although education issues are deemed more crucial. It may be thus deduced that increase in education, training and expertise can be a tool to address issues with performance.

The analysis shows the combined weight of barriers is 400% (normalized to 40) and the combined weight of the drivers is 300% (normalized to 30). By increasing attention to the drivers, via increased research and development, raising each to a 100%, the combined weight of the drivers will rise to 400% (normalized to 40)—assuming the barriers stay constant. In this scenario, the drivers will effectively cancel out the barriers.

With particular mention to the strategies proposed, our goal was to identify if there was sufficient information to suggest BIPV customization was a potential driver for BIPV adoption. To this end, the need for education related to design integration such as BIPV variety relating to technological choice, aesthetics, color, shape and size has been identified [25]. Improvements in product design with appeal to architects was mentioned as a potential solution; relating specifically to aesthetics [25,26,32,33], directly to customization and variety [25,26], and architectural integration [25,26,28] and innovation [30]. Thus, customization driven by variety, aesthetics and architectural integration has been identified as a potential driver of BIPV adoption. This justifies the need to further investigate BIPV customization studies and the validation of their potential to address these barriers as mentioned.

4. BIPV Façade Customization: Critical Review of Investigations

As architects are saddled with the responsibility of building design; it is pertinent to understand fully the opportunities provided by BIPV in order to communicate them effectively to clients [75]. It has been put forward that the success of the BIPV market will in part be determined by the availability of good customizability and convincing aesthetics [39]. It terms of the need, one aspect mentioned by [76] suggests that standardized products are often not applicable when retrofitting demands flexible dimensions, and custom products have better thermal performance than conventional products [77]. Another research posits that aesthetics, dimensional requirements focused on customization capacity and functionality ought to drive the requirements for the BIPV façade [78]. Thus, this calls for innovative approaches with custom-made products as some have huge potential for energy conservation and thermal comfort [79]. Most manufacturers provide custom-made BIPV services, such as the possibility to produce modules of various power output, form, glass serigraphy/printing and colors, as well as to change the cell arrangement and the glass surface (clear glass, prism, enameled) with different properties (i.e., glare reduction) and finishing [39]. This section however, focuses on BIPV façade customization from the perspective of research investigations to identify the potentials of custom BIPV already mentioned in this review. This perspective was chosen to detail the unbiased results of research experimentation without the inhibition of market performance or worthiness.

4.1. Methodological Approach

This section focuses on the details of the investigative analysis of the 25 selected papers on BIPV customization. As earlier stated, the result of the review was used to check the applicability of BIPV façade customization to address the barriers of standard BIPV. Four (4) aspects of review were selected which are Innovation & custom category, Customisation strategy, Architectural function and Research results. These describe various aspects of BIPV façade customization and form the framework for this evaluation. Figure 3 shows these related aspects as a research guide; with further explanation briefly presented following the figure to explain the definitions and state the importance of each aspect of the review.

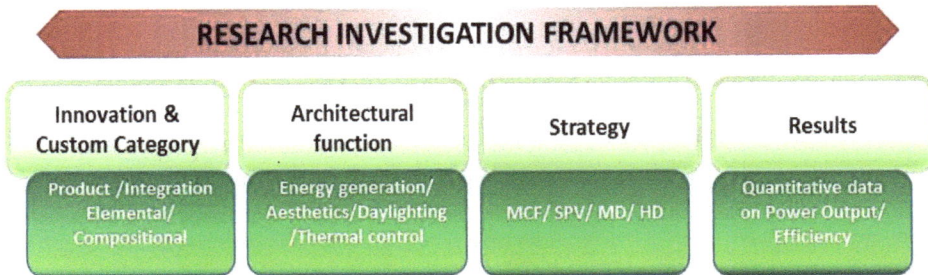

RESEARCH INVESTIGATION FRAMEWORK

Innovation & Custom Category	Architectural function	Strategy	Results
Product /Integration Elemental/ Compositional	Energy generation/ Aesthetics/Daylighting /Thermal control	MCF/ SPV/ MD/ HD	Quantitative data on Power Output/ Efficiency

Figure 3. Research investigation framework. Source: By Authors.

1. Innovation and Custom Category: Product/Integration/Elemental/Compositional

This establishes whether customization is the design of a new product or a developed system of integration. Next, the identification of the customization level with regards to the aspect of the BIPV façade for which parametric variation was investigated. In this regard, several other authors suggest that a sub-division of BIPV exists by its constitution, being the elements that make up the modules and the composition that makes up the façade [35,38,39,58,60].

- The Elemental Level: this represents the breakdown of a BIPV module into various components i.e., the solar cells, frame, glass and other protective layers; reflecting customization of cell or glass or layer type; colors or efficiencies.
- The Compositional Level: this represents the composition of the cells of the BIPV module (module-level), relating to cell spacing and the modules of the BIPV façade (façade-level), relating to tilt angle or spacing from wall for example.

2. Customization strategy: Standard/Custom/Module/Façade design manipulation

In each of the studies various strategies have been employed to customize the BIPV façade. This section helps to provide an inventory of various strategies, categorization for further analysis, possible requirements, limitations and challenges, as well as directions for possible improvements. These groups are;

- Systematic Parametric Variation (SPV): iterative parametric changes to reach an optimum goal
- Modification of Conventional Features (MCF): modification of conventional BIPV parts
- Enhanced Design Modularization (EDM): upgrade of BIPV façade types into unique modules
- Compositional Modification and Hybridization (CMH): combination of special materials with BIPV

3. Architectural function: energy generation/aesthetics/daylighting/thermal control

This section represents the stated, implied and potential functions of the custom BIPV façade in each study. It informs the specific custom function and provides justification to debate the sustained multi-functional advantage of BIPV. It also indicates if a connection exists between the customization category and level, and the potential architectural function.

4. Results: Power Output/Cell efficiency/Heating or Cooling loads

As the focus of this review is to validate the potential of BIPV customization as a solution to standard BIPV challenges, we also extracted the specific qualitative or quantitative data from the studies as made available. In some cases data on power output of BIPV façade output was provided which was;

- A comparison with a base case (standard BIPV);
- Hot climate results as representative of intense scenarios (where multiple climatic data was presented), or
- Highest output (where optimization based on parametric variation was investigated)

Table 4 details the investigated cases based on the research investigation framework explained above. It presents a concise summary to reflect how each study addressed BIPV customization to meet certain pre-defined objectives. All deductions from made from the experimental investigations were synthesized and analyzed in the discussion that follows the table.

Table 4. Research investigations on BIPV façade customization.

Reference	Country of Study (BIPV Location on Façade)	Objective	Deductions from Experimental Investigations				Research Results
			Custom Category/Class of Study	Customization Level Investigated	Strategy (Description)	Architectural Potential	
[51]	Taiwan (Wall)	Development and analysis of a full-colour PV module	Product/Design and fabrication	Elemental (Full-colour and monochromatic coloration of module parts)	Modification of conventional features (MCF) (Color image on backsheetglass with applied grayscale mask)	Energy generation; Aesthetics	Short Current density: 0–14% reduction; Cell efficiency : drop to max of 10%; Power: 14.2% reduction
[53]	Switzerland (External Device)	Design, fabrication and testing of an adaptive solar façade	Product & Integration/Design and fabrication; Performance Optimisation; Architectural Integration	Compositional @Façade-level (dynamic façade patterns and flexible tilt angle)	Enhanced design modularization (EDM) (Highly modular dynamic BIPV façade with a suitable support structure, tracking and control systems)	Energy generation; Thermal Control; Energy saving; Aesthetics	Power: 36% increase; Total energy savings: 31% increase; Energy consumption: 8.9% decrease; CO_2 offset: 15.3 kg CO_2-eq per year based on the European Union grid mix.
[80]	Austria (Glazing)	Plasmonic coloring on c-Si PV modules	Product/Design and fabrication	Elemental (Cell coloration)	MCF; (Silver film deposition on c-Si modules with Ag thermal annealing.)	Energy generation; Aesthetics	Short circuit current: average of 10.7% reduction; Open circuit voltage: average of 1.1% increase; Fill factor: average of 3.07% increase; Maximum Power/Efficiency: average of 8.3% reduction
[81]	South Korea (External Shading)	Application of layering effects to a BIPV façade	Product/Design and fabrication; Architectural Integration	Compositional @Module-level (Coloration of backsheet and Cell arrangement)	EDM (Layered effects to BIPV module: unique architectural finishing of glass sheets, coloration of backsheet with patterned cell arrangement.)	Energy generation; Daylighting; Aesthetics	Architectural layering and modularization approach enabled application and adaptation of the described effects specifically developed to meet unique requests from clients; No performance data was available.
		Modular retrofitting of a BIPV façade	Integration/Architectural Integration	Compositional @façade-level (tilt angle)	EDM (Modular retrofit and prototyping based on design of conventional façades)	Energy generation; Daylighting; Aesthetics	
[63]	Switzerland (Cladding)	Retrofitting of a prototype residential block with BIPV	Integration/Architectural Integration	Elemental (Cell transparency); Compositional @Façade-level (Module position)	EDM (adaptation of BIPV typologies to blend with conventional facade prototypes)	Energy generation; Thermal control; Daylighting; Retrofitting; Aesthetics	No extra complexity recorded in application of the method and façade construction. Qualitative assessment of interviewed professional adjudge that aesthetical aspects as positive .

Table 4. *Cont.*

Reference	Country of Study (BIPV Location on Façade)	Objective	Custom Category/Class of Study [*]	Customization Level Investigated	Strategy (Description)	Architectural Potential	Research Results
[82]	Korea (Window)	Colored a-Si:H transparent solar cells employing ultrathin transparent multi-layered electrodes.	Product/Performance and Optimization; Architectural Integration	Elemental (Electrode and Backsheet transparency, Colour variability)	EDM; systematic parametric variation (SPV); MCF (Transparent multi-layered electrodes (TMEs) with customizable coloration of optoelectronic controlling layer (OCL)	Energy generation; Aesthetics; Daylighting	Cell Efficiency: average of 6.36% at 23.5% average transmittance with TME @500–800 nm Ave Open circuit voltage: 0.8 V Ave Fill factor: 54.66%
[83]	China (Window & Double Skin Façade)	Comparison of energy performance between PV double skin façades and PV insulating glass units	Product/Design and fabrication; Performance and Optimization; Architectural Integration	Compositional @Module-level (Module Position, Air gap)	MCF; SPV; (Regulation of air gap)	Energy generation; Thermal control; Daylighting; Energy saving; Aesthetics	Ave. SHGCs: 0.152 (PV-DSF) and 0.238 (PV-IGU) Ave. U-value: 2.535 W/m²K (PV-DSF)and 2.281 W/m²K (PV-IGU) Conversion efficiency of PV-DSF is 1.8% better than PV-IGU Approx. power output: 0.01–0.3 kWh (PV-DSF); 0.01–0.32 kWh (PV-IGU) Energy Saving potential: 28.4% (PV-DSF) and 30% (PV-IGU)
[84]	China (Double Skin Façade)	Overall energy performance of an a-si based photovoltaic double-skin façade	Integration/Design and fabrication; Performance and optimization; Architectural Integration;	Compositional @Façade-level (Ventilation mode)	MCF; SPV; (Change in ventilated modes for PV-DSF)	Energy generation; Thermal control	Ave SHGC: 0.14 (Non-Ventilated), 0.15 (Naturally-Ventilated), 0.125 (Ventilated) U-value: 3.3 (Non-Ventilated), 3.7 (Naturally-Ventilated), 4.65 (Ventilated)
[85]	Switzerland (Window & Wall)	Performance investigation of selected BIPV façade types.	Product & Integration/Design and fabrication; Performance and optimization; Architectural Integration	Elemental (Cell technology; cell transparency); Compositional @Module-level & Façade-level (Module Position, Air gap, Tilt angle)	SPV; Variation of BIPV module position and ventilation mode	Energy generation Thermal control Daylighting Shading Aesthetics	Approx. power output: 3–11 kWh (c-Si @30°); 2.5–8 kWh (c-Si @90°); 0.6–2.1 kWh (a-Si @30°); 0.5–1.45 kWh (a-Si @90°); 0.8–2 kWh (a-Si @90°-ventilated)

Table 4. *Cont.*

Reference	Country of Study (BIPV Location on Façade)	Objective	Deductions from Experimental Investigations				
			Custom Category/Class of Study	Customization Level Investigated	Strategy (Description)	Architectural Potential	Research Results
[86]	China (Window)	Assessment of energy performance of semi-transparent PV insulating glass units	Product & Integration/Design and fabrication; Performance and optimization; Architectural Integration	Elemental (Backsheet); Compositional @Façade-level (Air gap)	MCF; SPV; (Variation of air gap and backsheet material)	Energy generation; Thermal control; Daylighting; Energy saving	Ave. PV temp: 23–42 °C Ave. Daylight illuminance: 0–360 lux Ave. Heat gain: -12.5–165 W/m^2 Power output @ air gap: 67.41 kWh @3 mm; 67.35 kWh @6 mm; 67.32 kWh @9 mm; 67.3 kWh @12 mm; 67.29 kWh @15 mm Power output @backsheet type: 67.32 kWh (Clear glass); 66.84 kWh (Low-e glass); 67.4 kWh (Low iron glass): 67.2 kWh (Tinted glass)
[87]	USA (Window)	Energy benefits from semi-transparent BIPV window and daylight-dimming systems	Product & Integration/Performance and Optimization; Architectural Integration	Elemental (Cell transparency and efficiency); Compositional @Façade-level (Orientation and WWR)	MCF; SPV; (Use of a DOE-2 based calculation algorithm simulations of parameterised vaules)	Energy generation; Thermal control; Daylighting; Shading; Energy savings	Power output range on south façade/month: 35.1–71.9 kWh @6.65 efficiency, 40% transparency, 48 W 46.4–95.4 kWh @8.82 efficiency, 20% transparency, 64 W 52.4–107.2 kWh @9.91 efficiency, 10% transparency, 72 W Approx. Annual Power output @WWR: 1165 kWh @10%; 3496 kWh @30%; 8157 @70%
[88]	Canada (Double Skin Façade)	Patterns of façade system design for enhanced energy performance	Product & Integration/Performance and Optimization; Architectural Integration	Compositional @Module-level (Module placement/arrangement)	MCF; SPV; EDM (Manipulation of planar geometry to induce increase in solar capture)	Energy generation; Thermal control; Aesthetics	Comparison with base case: Power Output: 20–80% increase Heating load: about 200% increase (worst case) Cooling load: about 52% reduction (best case) Peak electricity: peak spread of 4–5 h.
[55]	Pakistan (Wall)	Energy and Cost Saving of a Photovoltaic-Phase Change Material (PV-PCM) System	Product/Design and fabrication; Performance and Optimization; Cost	Elemental (Phase-change materials); Compositional @Module-level (Module design)	SPV;CHM; EDM (Passive cooling of BIPV with solid-liquid PCMs)	Energy generation; Thermal control	Temperature drop: 16% (PV PCM-1); 32.5% (PV PCM-2) Ave energy efficiency increase: 7% (PV PCM-1); 10% (PV PCM-2)

Table 4. *Cont.*

Reference	Country of Study (BIPV Location on Façade)	Deductions from Experimental Investigations				
		Objective	Custom Category/Class of Study	Customization Level Investigated	Strategy (Description)	Architectural Potential	Research Results
[89]	China (Wall)	Analysis and monitoring results of a BIPV façade using PV ceramic tiles	Product/Design and fabrication; Performance and Optimization; Architectural Integration	Compositional @Module-level (Module Position/Module Arrangement)	MCF(Replacement of module backsheet with ceramic tile)	Energy generation; Thermal control; Aesthetics	Ave. power output: 15–72 kWh (east); 15–65 kWh (West); 1–72 kWh (south); 0–18 kWh (North)
[90]	UAE (Double Skin Façade)	Performance and energetic improvements due to installation of semi-transparent PV cells	Product & Integration/Design and fabrication; Performance and Optimization; Architectural Integration	Compositional @Module-level & Façade-level (Number of glass layers, Ventilation mode)	MCF; SPV; (Application of alternate ventilation modes and number of glass layers)	Energy generation; Cladding	Sensible cooling energy need reduction: 1.5% (DSF forced vs. natural). 1.9% (Single Layer forced vs. natural) Peak power drop: 4% (DSF forced vs. natural). 2.3% (Single Layer forced vs. natural) Annual energy production increased by 2.5 (DSF) 6% (Single Layer)
[17]	USA (Ventilated Double Skin Façade)	Numerical investigation of the energy saving potential of a semi-transparent photovoltaic double-skin façade	Product & Integration/Performance and Optimization; Architectural Integration	Compositional @Façade-level (Air gap, Orientation)	MCF; SPV; EDM (Application of alternate air gaps and orientation in office room prototype room)	Energy generation; Thermal control; Daylighting; Shading	Approx. ave. electricity use: 300 kWh (100 mm); 310 kWh (200 mm); 285 kWh (400 mm); 270 kWh (600 mm) With 400 mm: Max power output range on south façade/month: 10.3 kWh (June)–20 kWh (November) Approx. Annual Energy output: 48 kWh/m² (East), 64 kWh/m² (South), 54 kWh/m² (West) Approx. cooling need: 18–270 MJ Approx. heating need: 0–35 MJ Ave. daylighting illuminance/month: 130–300 lux Observed 50% less net electricity that conventional glazing systems

Table 4. Cont.

Reference	Country of Study (BIPV Location on Façade)	Objective	Custom Category/Class of Study	Customization Level Investigated	Strategy (Description)	Architectural Potential	Research Results
				Deductions from Experimental Investigations			
[91]	Slovakia (Ventilated PV Façade)	Thermal Performance of a Ventilated PV Façade Coupled with PCM	Product & Integration/Design, Performance and Optimization; Architectural Integration	Compositional @Module-level (Module design—addition of PCM)	CMH; SPV (Hybridisation of BIPV with PCM layer)	Thermal control	PV temp decrease: up to 20 °C Peak temp. shift: more than 5 h
[92]	France (Ventilated PV Façade)	Experimental evaluation of a naturally ventilated PV double-skin building envelope in real operating conditions	Product & Integration/Design and fabrication; Performance and Optimization; Architectural Integration	Compositional @Module-level (Module arrangement)	MCF; SPV (Utilising the stack effect to cool a prototype pleated PV double façade)	Energy generation; Thermal control; Daylighting; Shading; Aesthetics	Approx. Peak power output per plane: 165 kW (Bloc1); 200 kW (Bloc2); 210 kW (Bloc3) Prismatic configuration was chosen to compensate for façade azimuth—overshadowing in part; improvement in electrical performance by a more favorable orientation of solar cells
[93]	UAE (Window Blinds)	Energy, Cooling and Cost analysis of BIPV blind system	Product & Integration/Design and fabrication; Performance and Optimization; Architectural Integration; Cost	Elemental (cell technology); Compositional (Module position)	EDM; SPV (Prototyping based on conventional façade design component)	Energy generation; Thermal control; Cost issues	Ave. power output: 41.55 kWh/m^2 (c-Si); 43.22 kWh/m^2 (a-Si) Cooling load Energy Saved: 7.11 kWh/m^2 (c-Si); 6.89 kWh/m^2 (a-Si)
[64]	China (PV-Blinds in Double Skin Façade)	Comparative study on thermal performance evaluation of a new double skin façade system integrated with photovoltaic blinds	Product & Integration/Design and fabrication; Performance and Optimization; Architectural Integration	Compositional @Module-level (Module Position/Tilt angle)	SPV (Experimentation on different system ventilation modes and blind parameters)	Energy generation; Thermal control; Daylighting	Approx. SGHC peak (@4.5 cm spacing): 0.75 (30°); 0.95 (45°); 0.97 (60°); (based on ventilation mode) 0.499 (Mechanical) 0.531 (Natural) About 12.16% and 25.57% compared with reference DSF cases

Table 4. *Cont.*

Reference	Country of Study (BIPV Location on Façade)	Objective	Deductions from Experimental Investigations				
			Custom Category/Class of Study	Customization Level Investigated	Strategy (Description)	Architectural Potential	Research Results
[94]	China (Ventilated Double Skin Façade)	Thermal performance of a photovoltaic wall mounted on a multi-layer façade	Integration/Performance and Optimization; Architectural Integration	Compositional @Façade-level (ventilation mode)	SPV (Mathematical modelling and variation of ventilation modes)	Energy generation; Thermal control	Ave SHGC: 0.14 (Non-Ventilated), 0.15 (Naturally-Ventilated), 0.125 (Ventilated) U-value: 3.3 (Non-Ventilated), 3.7 (Naturally-Ventilated), 4.65 (Ventilated)
[52]	(Glazing)	Aesthetic improvement of PV for Building integration Encapsulants	Product/Design and fabrication; Architectural Integration	Elemental (Encapsulant material)	MCF (Coloration of encapsulant material using florescence dyes)	Energy generation; Aesthetics	Power output increase: 2.0 % (Clear Sylgard 184); 2.5% (Red 100 ppm Lumogen dye in Sylgard 184)
[95]	Italy (Glassblocks)	Evaluation of prototype BIPV optical performance	Product/Design and Fabrication; Performance optimization	Compositional @Module-level (position of solar cells)	CMH; EDM (Prototyping based on conventional façade design component)	Energy generation; Aesthetics; Daylighting; Thermal Control	Power output reductions: 19.67% (DSSC Part of Surface); 6.01 % (All of Surface); 54.09% (Interior of Surface); 69.94% (Middle of Block)
[96]	Netherlands (Wall)	Aesthetics preservation BIPV façade using Zigzag geometry	Product/Design and Fabrication; Architectural integration	Elemental (colour of reflector layer); Compositional @Façade-level (tilt angle)	EDM; MCF; SPV (Concealment of PV via zigzag geometry to enhance solar capture)	Energy generation; Aesthetics	Monthly Power output: 28.6 kWh (Grey), 30.7 kWh (White) Performance ratio increase (ref. vertical panels): 43.75% (Grey), 53.75% (White)

4.2. Assessment of BIPV Customization Parameters

Table 4 shows that all the studies reviewed give focus to energy generation and architectural integration; most also focus on performance and optimization of the BIPV façade; few focus on cost and environmental issues. This is reminiscent of the general fact that a BIPV façade is primarily a building element with energy producing capability. Thus, its integration and optimization of its performance are significantly important. The country of study and BIPV façade type highlight the potentiality in a variety of countries and application in building location. The variety of objective and approach in the various studies was expected, and provided a broad spectrum to carry out the review. However, in order to provide a sensible analysis, categorization was done at each stage without bias to the original intent of the researchers.

4.2.1. Innovation and Custom Category

Statistically, nine of the cases focused on design of a custom BIPV product [51,52,55,80–83,89,95], only four focused on a customization in the integration process [63,81,84,94], while 12 combined both product and integration concerns in their research [17,53,63,64,85–88,90–93]. This suggests that most custom BIPV façade products are designed with attention on the potential for proper architectural integration. Each approach is uniquely different, yet they meet the same goals of energy generation, aesthetics, and daylighting or thermal control. It is important to consolidate at this point the fact that conventional BIPV façades can provide some of these gains along with energy generation. However, these customized BIPV have the potential to out-perform standard types based on pre-design specifications and functionally-driven objectives, which emphasize these other benefits.

Regarding the customization level, four were purely elemental [51,52,80,82], eight were compositional at the module-level [64,81,83,88,89,91,92,95], and five were compositional at the façade-level [17,53,81,84,94]; eight studies combined all of the levels [55,82,85–87,90,93,96]. Comparing elemental versus compositional level, studies can be more easily carried out using conventional PV modules without the requirement of a custom-designed module. As this will require less time to fabricate the test specimens, it is probable that compositional studies are thus preferred, and were thus more numerous. However, using conventional modules for customization suggests innovative and creative applications.

4.2.2. Customization Strategy

The studies showed varying levels of complexity in the strategies used to achieve the objective of customization. It is clear from the examples that this was achieved by an interdisciplinary approach to BIPV product design. It therefore suggests that the accomplishment of custom BIPV modules requires input across several disciplines. While this may be more demanding and expensive, it creates the opportunity for greater novelty and innovative ideas. Enhanced flexibility and variety was noticed in the strategic approach applied in custom BIPV integration studies.

In the investigated cases, three were SPV [64,85,94], four were MCF [51,52,80,89], four were EDM [53,63,81], and 14 combined two or more strategies [17,55,82–84,86–88,90–93,95,96]. Clear evidence thus presents a combination of various strategies is required to achieve BIPV façade customization. In the cases of combined strategies, most of the studies addressed customization at both an elemental and compositional level, reflecting a holistic approach. Furthermore, most of the studies in this class were carried out to address aesthetic or thermal control objectives. Deductively therefore, the combined strategy approach is preferred for BIPV façade customization as it covers various multi-dimensional issues in the design.

4.2.3. Architectural Function

All the studied cases showed interest in energy generation potential of the custom BIPV, but to varying degrees. With regards to the added functions, [17] studies addressed thermal control [17,53,63,83–89,91–95], [15] addressed aesthetics [17] and, [11] addressed daylighting functionality based on research objective or cell type selection as all a-Si applications permit some degree of light transmission [7,63,64,82,83,85–87,92,95]. Also, three addressed energy savings [53,83,87], four specifically on shading [85–87,92], and one on cladding [90] and cost [93].

The review shows that although all focused on energy generation of the custom BIPV façade, all focused also on at least one or more added function. More than half of the studies were on thermal control—in terms of added BIPV function; proving capture and reuse of PV thermal energy, as well as reduction of direct solar radiation to the interior. A sizeable number of the studies show a connection with the goal of improving the thermal control or aesthetic appeal of the product. It suggests that customization of BIPV products is in some way primarily driven by these two objectives. We observed that the architectural functions were achieved by all the classified strategies. The import of this finding is that categorization developed for this review of customization strategies for BIPV façades is justifiable, flexible and versatile in applicability.

4.2.4. Results

Performance data from the studies were varied and not reflective of comparisons with a conventional BIPV or a reference case in most cases. Were available, a 4–70% reduction related to power output was observed [51,80,90,95] and a 2–80% increase [52,53,55,96]. It is important to observe that these studies used different strategies with different reference cases. As these studies were also in different climates it is not possible to make a general conclusion on these results. They are however representative of the fact that BIPV façade customization has potentials for performance improvements or otherwise based on the design and specifications

Of the studies related to thermal control with reference cases, 3 showed improvements in relation to power output while 1 showed reductions. Of the studies related aesthetics with reference cases, 4 showed improvements in relation to power output while 2 showed reductions. This suggests the in comparative situations, the process of customization can enhance thermal control and aesthetics with satisfactory performance related to power output.

5. Challenges and Future Prospects

Several challenges exist with the concept of BIPV customization in general and specific terms and several studies have outlined these barriers [24–34,97]. Firstly, BIPV itself is still in a technological developmental phase. Its full potential is yet to be maximized and studies argue that there are still design codes and standards for application that are not full developed [24,28,30]. This review has further brought to light the vast variations in strategies and approaches with BIPV façade customization. Developing a framework for analysis is thus potentially challenging and requires certain generalizations.

Specific to BIPV façade customization objectives, thermal control and aesthetics were identified as the most frequently studied. However, the several of strategies used in both cases required special manufacturing processes which are not yet standardized on a large scale. Thus, problems with cost, machinery, and standards exist. This scenario is worsened by the identified gap between the PV and building industry [29,30,32]; as the lack of willingness to adopt new technology can be a drawback for custom applications. Further research is required to standardize the assessment of custom BIPV and develop a model for evaluation of strategies. This review intended to identify the potentiality of BIPV customization but does not answer questions related to climatic or regional applications. Customization in relation to cost issues is another area that presents further research potential to yield clear evaluative data. The cost and efficiency of a BIPV system can be lowered by reducing PV module and component manufacturing costs, improving PV and other component efficiencies, and understanding whole life cycle costs [21,98] in relation to local factors and the context when used a building skin [99].

A comparative analysis between research driven BIPV customization and commercial custom BIPV products is also required to reflect the quantitative and qualitative dimensions of the variations in performance and perception. The main bottleneck discovered during a BIPV study conducted in an European research project, was in the ability to communicate this enhanced value and the new possibilities to customers and thus justify the higher cost -generally an increment around 20% [66]. Thus, custom BIPV potentials require proper communication of potentials to both the public and professionals.

6. Conclusions

This review strategically raises a theoretical background for a renewed focus on BIPV customization. It is clear that there are several experimental studies which engage in this strategy at one level or the other. Our findings indicate that BIPV façade customization can be carried out with significant advantages which include:

1. Flexibility and applicability at an elemental and compositional level
2. Versatility in development of both custom BIPV products and custom BIPV integration schemes
3. Multiple type strategies in single or combined scenarios can be used to achieve objectives
4. Increase in power output and performance is possible in a range of 2–80% based on design
5. Although, reduction in power output and performance occurs also at a range of 4–70% based on design

In summary, we conclude that BIPV façade customization can address some of the barriers with conventional BIPV façades relating to product efficiency and aesthetic design. It can also be a driver of enhanced innovative product design for architectural integration. The extensive research and global interest in BIPV over the last one decade is not likely to abate. Areas such as daylighting, self-cleaning PV glazing, aesthetics using color, form or shapes, concentrating BIPV, perovskite-based solar cells and solar trees are some of the emerging areas [36,45,100–103]. With shifting policies, government tariffs and policy changes, it will also be interesting to investigate the possibility of using demonstration projects in certain regions as a push for BIPV-wide acceptance. Such projects will be opportunities to communicate the significant benefits of BIPV customisation and advance its adoption.

Acknowledgments: The authors gratefully acknowledge financial support from the United Arab Emirates University towards this review as part of an on-going doctoral research.

Author Contributions: Daniel Efurosibina Attoye developed the concept and framework for investigation, carried out data collection and wrote the initial manuscript. Kheira Tabet Aoul and Ahmed Hassan verified the scholastic depth of the paper, carried out a critical review of the structure and content, and also supervised the analysis and findings of this work. All authors discussed the results and contributed to the final manuscript.

Conflicts of Interest: The authors declare no conflict of interest.

References

1. Nejat, P.; Jomehzadeh, F.; Taheri, M.M.; Gohari, M.; Majid, M.Z.A. A global review of energy consumption, CO_2 emissions and policy in the residential sector (with an overview of the top ten CO_2 emitting countries). *Renew. Sustain. Energy Rev.* **2015**, *43*, 843–862. [CrossRef]
2. World Energy Council (WEC). *World Energy Resources 2013 Survey*; World Energy Council: London, UK, 2013.
3. De la Cruz-Lovera, C.; Perea-Moreno, A.J.; de la Cruz-Fernández, J.L.; Alvarez-Bermejo, J.A.; Manzano-Agugliaro, F. Worldwide Research on Energy Efficiency and Sustainability in Public Buildings. *Sustainability* **2017**, *9*, 1294. [CrossRef]
4. Woodcock, J.; Edwards, P.; Tonne, C.; Armstrong, B.G.; Ashiru, O.; Banister, D.; Beevers, S.; Chalabi, Z.; Chowdhury, Z.; Cohen, A.; et al. Public health benefits of strategies to reduce greenhouse-gas emissions: Urban land transport. *Lancet* **2009**, *374*, 1930–1943. [CrossRef]
5. Adewuyi, A.O.; Awodumi, O.B. Renewable and non-renewable energy-growth-emissions linkages: Review of emerging trends with policy implications. *Renew. Sustain. Energy Rev.* **2017**, *69*, 275–291. [CrossRef]
6. Wang, F.; Wang, C.; Su, Y.; Jin, L.; Wang, Y.; Zhang, X. Decomposition Analysis of Carbon Emission Factors from Energy Consumption in Guangdong Province from 1990 to 2014. *Sustainability* **2017**, *9*, 274. [CrossRef]

7. Rodhe, H. A comparison of the contribution of various gases to the greenhouse effect. *Science* **1990**, *248*, 1217. [CrossRef] [PubMed]

8. Lashof, D.A.; Ahuja, D.R. Relative contributions of greenhouse gas emissions to global warming. *Nature* **1990**, *344*, 529–531. [CrossRef]

9. Jiang, W.; Liu, J.; Liu, X. Impact of carbon quota allocation mechanism on emissions trading: An agent-based simulation. *Sustainability* **2016**, *8*, 826. [CrossRef]

10. Camanzi, L.; Alikadic, A.; Compagnoni, L.; Merloni, E. The impact of greenhouse gas emissions in the EU food chain: A quantitative and economic assessment using an environmentally extended input-output approach. *J. Clean. Prod.* **2017**, *157*, 168–176. [CrossRef]

11. Intergovernmental Panel on Climate Change (IPCC). *Climate Change 2007: Synthesis Report*; Intergovernmental Panel on Climate Change: Geneva, Switzerland, 2007; pp. 45–54.

12. Liu, W.Y.; Lin, C.C.; Chiu, C.R.; Tsao, Y.S.; Wang, Q. Minimizing the carbon footprint for the time-dependent heterogeneous-fleet vehicle routing problem with alternative paths. *Sustainability* **2014**, *6*, 4658–4684. [CrossRef]

13. Knera, D.; Knera, D.; Heim, D.; Heim, D. Application of a BIPV to cover net energy use of the adjacent office room. *Manag. Environ. Qual. Int. J.* **2016**, *27*, 649–662. [CrossRef]

14. Evola, G.; Margani, G. Renovation of apartment blocks with BIPV: Energy and economic evaluation in temperate climate. *Energy Build.* **2016**, *130*, 794–810. [CrossRef]

15. Jayathissa, P.; Luzzatto, M.; Schmidli, J.; Hofer, J.; Nagy, Z.; Schlueter, A. Optimising building net energy demand with dynamic BIPV shading. *Appl. Energy* **2017**, *202*, 726–735. [CrossRef]

16. Wong, P.W.; Shimoda, Y.; Nonaka, M.; Inoue, M.; Mizuno, M. Semi-transparent PV: Thermal performance, power generation, daylight modelling and energy saving potential in a residential application. *Renew. Energy* **2008**, *33*, 1024–1036. [CrossRef]

17. Peng, J.; Curcija, D.C.; Lu, L.; Selkowitz, S.E.; Yang, H.; Zhang, W. Numerical investigation of the energy saving potential of a semi-transparent photovoltaic double-skin facade in a cool-summer Mediterranean climate. *Appl. Energy* **2016**, *165*, 345–356. [CrossRef]

18. Bayoumi, M. Impacts of window opening grade on improving the energy efficiency of a façade in hot climates. *Build. Environ.* **2017**, *119*, 31–43. [CrossRef]

19. Elinwa, U.K.; Radmehr, M.; Ogbeba, J.E. Alternative Energy Solutions Using BIPV in Apartment Buildings of Developing Countries: A Case Study of North Cyprus. *Sustainability* **2017**, *9*, 1414. [CrossRef]

20. Song, A.; Lu, L.; Liu, Z.; Wong, M.S. A Study of Incentive Policies for Building-Integrated Photovoltaic Technology in Hong Kong. *Sustainability* **2016**, *8*, 769. [CrossRef]

21. Norton, B.; Eames, P.C.; Mallick, T.K.; Huang, M.J.; McCormack, S.J.; Mondol, J.D.; Yohanis, Y.G. Enhancing the performance of building integrated photovoltaics. *Sol. Energy* **2011**, *85*, 1629–1664. [CrossRef]

22. Agathokleous, R.A.; Kalogirou, S.A. Double skin facades (DSF) and building integrated photovoltaics (BIPV): A review of configurations and heat transfer characteristics. *Renew. Energy* **2016**, *89*, 743–756. [CrossRef]

23. Zhang, W.; Lu, L.; Peng, J. Evaluation of potential benefits of solar photovoltaic shadings in Hong Kong. *Energy* **2017**. [CrossRef]

24. Ritzen, M.; Reijenga, T.; El Gammal, A.; Warneryd, M.; Sprenger, W.; Rose-Wilson, H.; Payet, J.; Morreau, V.; Boddaert, S. IEA-PVPS Task 15: Enabling Framework for BIPV Acceleration. (IEA-PVPS). In Proceedings of the 48th IEA PVPS Executive Commitee Meeting, Vienna, Austria, 16 November 2016.

25. Tabakovic, M.; Fechner, H.; Van Sark, W.; Louwen, A.; Georghiou, G.; Makrides, G.; Loucaidou, E.; Ioannidou, M.; Weiss, I.; Arancon, S.; et al. Status and outlook for building integrated photovoltaics (BIPV) in relation to educational needs in the BIPV sector. *Energy Procedia* **2017**, *111*, 993–999. [CrossRef]

26. Prieto, A.; Knaack, U.; Auer, T.; Klein, T. Solar façades-Main barriers for widespread façade integration of solar technologies. *J. Façade Des. Eng.* **2017**, *5*, 51–62. [CrossRef]

27. Goh, K.C.; Goh, H.H.; Yap, A.B.K.; Masrom, M.A.N.; Mohamed, S. Barriers and drivers of Malaysian BIPV application: Perspective of developers. *Procedia Eng.* **2017**, *180*, 1585–1595. [CrossRef]

28. Yang, R.J.; Zou, P.X. Building integrated photovoltaics (BIPV): Costs, benefits, risks, barriers and improvement strategy. *Int. J. Constr. Manag.* **2016**, *16*, 39–53. [CrossRef]

29. Karakaya, E.; Sriwannawit, P. Barriers to the adoption of photovoltaic systems: The state of the art. *Renew. Sustain. Energy Rev.* **2015**, *49*, 60–66. [CrossRef]

30. Yang, R.J. Overcoming technical barriers and risks in the application of building integrated photovoltaics (BIPV): Hardware and software strategies. *Autom. Constr.* **2015**, *51*, 92–102. [CrossRef]
31. Azadian, F.; Radzi, M.A.M. A general approach toward building integrated photovoltaic systems and its implementation barriers: A review. *Renew. Sustain. Energy Rev.* **2013**, *22*, 527–538. [CrossRef]
32. Koinegg, J.; Brudermann, T.; Posch, A.; Mrotzek, M. "It Would Be a Shame if We Did Not Take Advantage of the Spirit of the Times". An Analysis of Prospects and Barriers of Building Integrated Photovoltaics. *GAIA Ecol. Perspect. Sci. Soc.* **2013**, *22*, 39–45. [CrossRef]
33. Probst, M.M.; Roecker, C. Criteria for architectural integration of active solar systems IEA Task 41, Subtask A. *Energy Procedia* **2012**, *30*, 1195–1204. [CrossRef]
34. Taleb, H.M.; Pitts, A.C. The potential to exploit use of building-integrated photovoltaics in countries of the Gulf Cooperation Council. *Renew. Energy* **2009**, *34*, 1092–1099. [CrossRef]
35. Tripathy, M.; Sadhu, P.K.; Panda, S.K. A critical review on building integrated photovoltaic products and their applications. *Renew. Sustain. Energy Rev.* **2016**, *61*, 451–465. [CrossRef]
36. Jelle, B.P. Building integrated photovoltaics: A concise description of the current state of the art and possible research pathways. *Energies* **2016**, *9*, 21. [CrossRef]
37. Heinstein, P.; Ballif, C.; Perret-Aebi, L.E. Building integrated photovoltaics (BIPV): Review, potentials, barriers and myths. *Green* **2013**, *3*, 125–156. [CrossRef]
38. Cerón, I.; Caamaño-Martín, E.; Neila, F.J. 'State-of-the-art' of building integrated photovoltaic products. *Renew. Energy* **2013**, *58*, 127–133. [CrossRef]
39. Bonomo, P.; Chatzipanagi, A.; Frontini, F. Overview and analysis of current BIPV products: New criteria for supporting the technological transfer in the building sector. *VITRUVIO Int. J. Archit. Technol. Sustain.* **2015**, 67–85. [CrossRef]
40. Shukla, A.K.; Sudhakar, K.; Baredar, P. Recent advancement in BIPV product technologies: A review. *Energy Build.* **2017**. [CrossRef]
41. Ibraheem, Y.; Farr, E.R.; Piroozfar, P.A. Embedding passive intelligence into building envelopes: A review of the state-of-the-art in integrated photovoltaic shading devices. *Energy Procedia* **2017**, *111*, 964–973. [CrossRef]
42. Shukla, A.K.; Sudhakar, K.; Baredar, P. A comprehensive review on design of building integrated photovoltaic system. *Energy Build.* **2016**, *128*, 99–110. [CrossRef]
43. Skandalos, N.; Karamanis, D. PV glazing technologies. *Renew. Sustain. Energy Rev.* **2015**, *49*, 306–322. [CrossRef]
44. Jelle, B.P.; Breivik, C.; Røkenes, H.D. Building integrated photovoltaic products: A state-of-the-art review and future research opportunities. *Sol. Energy Mater. Sol. Cells* **2012**, *100*, 69–96. [CrossRef]
45. Jelle, B.P.; Breivik, C. The path to the building integrated photovoltaics of tomorrow. *Energy Procedia* **2012**, *20*, 78–87. [CrossRef]
46. Hammond, G.P.; Harajli, H.A.; Jones, C.I.; Winnett, A.B. Whole systems appraisal of a UK Building Integrated Photovoltaic (BIPV) system: Energy, environmental, and economic evaluations. *Energy Policy* **2012**, *40*, 219–230. [CrossRef]
47. Scognamiglio, A.; Røstvik, H.N. Photovoltaics and zero energy buildings: A new opportunity and challenge for design. *Prog. Photovolt. Res. Appl.* **2013**, *21*, 1319–1336. [CrossRef]
48. Scognamiglio, A.; Farkas, K.; Frontini, F.; Maturi, L. Architectural quality and photovoltaic products. In Proceedings of the 27th European Photovoltaic Solar Energy Conference and Exhibition (EU PVSEC), Frankfurt, Germany, 24–28 September 2012; pp. 24–28.
49. Stevenson, A. (Ed.) *Oxford dictionary of English*; Oxford University Press: New York, NY, USA, 2010.
50. Pine, B.J. Mass customizing products and services. *Plan. Rev.* **1993**, *21*, 6–55. [CrossRef]
51. Lien, S.Y. Artist Photovoltaic Modules. *Energies* **2016**, *9*, 551. [CrossRef]
52. Hardy, D.; Kerrouche, A.; Roaf, S.C.; Richards, B.S. Improving the Aesthetics of Photovoltaics through Use of Coloured Encapsulants. In Proceedings of the PLEA 2013—29th Conference, Sustainable Architecture for a Renewable Future, Munich, Germany, 10–12 September 2013.
53. Nagy, Z.; Svetozarevic, B.; Jayathissa, P.; Begle, M.; Hofer, J.; Lydon, G.; Willmann, A.; Schlueter, A. The adaptive solar facade: From concept to prototypes. *Front. Archit. Res.* **2016**, *5*, 143–156. [CrossRef]
54. Keller, A.F. Recharging the Facade: Designing and Constructing Novel BIPV Assemblies. Ph.D. Thesis, Massachusetts Institute of Technology, Cambridge, MA, USA, 2013.

55. Hasan, A.; McCormack, S.J.; Huang, M.J.; Norton, B. Energy and cost saving of a photovoltaic-phase change materials (PV-PCM) system through temperature regulation and performance enhancement of photovoltaics. *Energies* **2014**, *7*, 1318–1331. [CrossRef]

56. Hasan, A.; McCormack, S.J.; Huang, M.J.; Sarwar, J.; Norton, B. Increased photovoltaic performance through temperature regulation by phase change materials: Materials comparison in different climates. *Sol. Energy* **2015**, *115*, 264–276. [CrossRef]

57. Hasan, A.; Sarwar, J.; Alnoman, H.; Abdelbaqi, S. Yearly energy performance of a photovoltaic-phase change material (PV-PCM) system in hot climate. *Sol. Energy* **2017**, *146*, 417–429. [CrossRef]

58. Baum, R. Architectural integration of light-transmissive photovoltaic (LTPV). In Proceedings of the 26th European Photovoltaic Solar Energy Conference and Exhibition (EU PVSEC), Hamburg, Germany, 5–9 September 2011; pp. 5–9.

59. Munari Probst, M.C.; Roecker, C.; Frontini, F.; Scognamiglio, A.; Farkas, K.; Maturi, L.; Zanetti, I. Solar Energy Systems in Architecture-integration criteria and guidelines. In *International Energy Agency Solar Heating and Cooling Programme*; Probst, M., Cristina, M., Roecker, C., Eds.; International Energy Agency: Paris, France, 2013.

60. Farkas, K.; Frontini, F.; Maturi, L.; Munari Probst, M.C.; Roecker, C.; Scognamiglio, A. *Designing Photovoltaic Systems for Architectural Integration*; Farkas, K., Ed.; International Energy Agency: Paris, France, 2013.

61. Montoro, D.F.; Vanbuggenhout, P.; Ciesielska, J. Building Integrated Photovoltaics: An overview of the existing products and their fields of application. In *Report Prepared in the Framework of the European Funded Project*; SUNRISE: Saskatoon, Canada, 2011.

62. Thomas, R. (Ed.) *Photovoltaics and Architecture*; Taylor & Francis: Didcot, UK, 2003.

63. Clua Longas, A.; Lufkin, S.; Rey, E. Towards Advanced Active Façades: Analysis of façade requirements and development of an innovative construction system. In Proceedings of the PLEA 2017, Edinburgh, UK, 3–5 July 2017; Volume 1, pp. 192–199.

64. Luo, Y.; Zhang, L.; Wang, X.; Xie, L.; Liu, Z.; Wu, J.; Zhang, Y.; He, X. A comparative study on thermal performance evaluation of a new double skin façade system integrated with photovoltaic blinds. *Appl. Energy* **2017**, *199*, 281–293. [CrossRef]

65. Tabriz, S.N.; Fard, F.; Partovi, N. Review of architectural day lighting analysis of photovoltaic panels of BIPV with zero energy emission approach. *Res. J. Appl. Sci.* **2016**, *11*, 735–741.

66. Pagliaro, M.; Ciriminna, R.; Palmisano, G. BIPV: Merging the photovoltaic with the construction industry. *Prog. Photovolt. Res. Appl.* **2010**, *18*, 61–72. [CrossRef]

67. Oliver, M.; Jackson, T. Energy and economic evaluation of building-integrated photovoltaics. *Energy* **2001**, *26*, 431–439. [CrossRef]

68. Bakos, G.C.; Soursos, M.; Tsagas, N.F. Technoeconomic assessment of a building-integrated PV system for electrical energy saving in residential sector. *Energy Build.* **2003**, *35*, 757–762. [CrossRef]

69. Sharples, S.; Radhi, H. Assessing the technical and economic performance of building integrated photovoltaics and their value to the GCC society. *Renew. Energy* **2013**, *55*, 150–159. [CrossRef]

70. Byrnes, L.; Brown, C.; Foster, J.; Wagner, L.D. Australian renewable energy policy: Barriers and challenges. *Renew. Energy* **2013**, *60*, 711–721. [CrossRef]

71. Morris, S. *Improving Energy Efficient, Sustainable Building Design and Construction in Australia—Learning from Europe*; ISS Institute: Carlton, Australia, 2013.

72. Abdullah, A.S.; Abdullah, M.P.; Hassan, M.Y.; Hussin, F. Renewable energy cost-benefit analysis under Malaysian feed-in-tariff. In Proceedings of the 2012 IEEE Student Conference on Research and Development (SCOReD), Pulau Pinang, Malaysia, 5–6 December 2012; pp. 160–165.

73. Sintov, N.D.; Schultz, P. Adjustable Green Defaults Can Help Make Smart Homes More Sustainable. *Sustainability* **2017**, *9*, 622. [CrossRef]

74. Martyn, A.S. Some problems in managing complex development projects. *Long Range Plann.* **1975**, *8*, 13–26. [CrossRef]

75. Attoye, D.E.; Tabet Aoul, K.A.; Hassan, A. Potentials and Benefits of Building Integrated Photovoltaics. In Proceedings of the United Arab Emirates Graduate Student Conference (UAEGSRC), Al Ain, UAE, 27–28 April 2016.

76. Jahanara, A. Strategy towards Solar Architecture by Photovoltaic for Building Integration. Ph.D. Thesis, Eastern Mediterranean University (EMU), Famagusta, Turkey, 2013.

77. Wu, Y.; Krishnan, P.; Liya, E.Y.; Zhang, M.H. Using lightweight cement composite and photocatalytic coating to reduce cooling energy consumption of buildings. *Constr. Build. Mater.* **2017**, *145*, 555–564. [CrossRef]
78. Biyik, E.; Araz, M.; Hepbasli, A.; Shahrestani, M.; Yao, R.; Shao, L.; Essah, E.; Oliveira, A.C.; del Caño, T.; Rico, E.; et al. A key review of building integrated photovoltaic (BIPV) systems. *Eng. Sci. Technol. Int. J.* **2017**. [CrossRef]
79. Tak, S.; Woo, S.; Park, J.; Park, S. Effect of the Changeable Organic Semi-Transparent Solar Cell Window on Building Energy Efficiency and User Comfort. *Sustainability* **2017**, *9*, 950. [CrossRef]
80. Peharz, G.; Berger, K.; Kubicek, B.; Aichinger, M.; Grobbauer, M.; Gratzer, J.; Nemitz, W.; Großschädl, B.; Auer, C.; Prietl, C.; et al. Application of plasmonic coloring for making building integrated PV modules comprising of green solar cells. *Renew. Energy* **2017**, *109*, 542–550. [CrossRef]
81. Van Berkel, T.; Minderhoud, T.; Piber, A.; Gijzen, G. Design Innovation from PV-Module to Building Envelope: Architectural Layering and Non Apparent Repetition. In Proceedings of the 29th European Photovoltaic Solar Energy Conference and Exhibition, Amsterdam, The Netherlands, 22–26 September 2014; pp. 366–372.
82. Lim, J.W.; Kim, G.; Shin, M.; Yun, S.J. Colored a-Si: H transparent solar cells employing ultrathin transparent multi-layered electrodes. *Sol. Energy Mater. Sol. Cells* **2017**, *163*, 164–169. [CrossRef]
83. Wang, M.; Peng, J.; Li, N.; Yang, H.; Wang, C.; Li, X.; Lu, T. Comparison of energy performance between PV double skin facades and PV insulating glass units. *Appl. Energy* **2017**, *194*, 148–160. [CrossRef]
84. Peng, J.; Lu, L.; Yang, H.; Sing, A.; Ma, T. 182: Investigation on the overall energy performance of an a-si based photovoltaic double-skin facade in Hong Kong. In Proceedings of the Sustainable Energy for a Resilient Future: The 14th International Conference on Sustainable Energy Technologies, Nottingham, UK, 25–27 August 2015; Rodrigues, L., Ed.; University of Nottingham: Nottingham, UK, 2015; Volume 1, pp. 263–271.
85. Chatzipanagi, A.; Frontini, F.; Virtuani, A. BIPV-temp: A demonstrative Building Integrated Photovoltaic installation. *Appl. Energy* **2016**, *173*, 1–2. [CrossRef]
86. Wang, M.; Peng, J.; Li, N.; Lu, L.; Ma, T.; Yang, H. Assessment of energy performance of semi-transparent PV insulating glass units using a validated simulation model. *Energy* **2016**, *112*, 538–548. [CrossRef]
87. Do, S.L.; Shin, M.; Baltazar, J.C.; Kim, J. Energy benefits from semi-transparent BIPV window and daylight-dimming systems for IECC code-compliance residential buildings in hot and humid climates. *Sol. Energy* **2017**, *155*, 291–303. [CrossRef]
88. Hachem, C.; Elsayed, M. Patterns of façade system design for enhanced energy performance of multistory buildings. *Energy Build.* **2016**, *130*, 366–377. [CrossRef]
89. Huang, Y.C.; Lee, S.K.; Chan, C.C.; Wang, S.J. Full-scale evaluation of fire-resistant building integrated photovoltaic systems with different installation positions of junction boxes. *Indoor Built Environ.* **2017**. [CrossRef]
90. Elarga, H.; Zarrella, A.; De Carli, M. Dynamic energy evaluation and glazing layers optimization of facade building with innovative integration of PV modules. *Energy Build.* **2016**, *111*, 468–478. [CrossRef]
91. Curpek, J.; Hraska, J. Simulation Study on Thermal Performance of a Ventilated PV Façade Coupled with PCM. In *Applied Mechanics and Materials*; Trans Tech Publications: Zürich, Switzerland, 2017; Volume 861, pp. 167–174.
92. Gaillard, L.; Giroux-Julien, S.; Ménézo, C.; Pabiou, H. Experimental evaluation of a naturally ventilated PV double-skin building envelope in real operating conditions. *Sol. Energy* **2014**, *103*, 223–441. [CrossRef]
93. Bahr, W. A comprehensive assessment methodology of the building integrated photovoltaic blind system. *Energy Build.* **2014**, *82*, 703–708. [CrossRef]
94. Peng, J.; Lu, L.; Yang, H.; Han, J. Investigation on the annual thermal performance of a photovoltaic wall mounted on a multi-layer façade. *Appl. Energy* **2013**, *112*, 646–656. [CrossRef]
95. Buscemi, A.; Calabrò, C.; Corrao, R.; Di Maggio, M.S.; Morini, M.; Pastore, L. Optical Performance Evaluation of DSSC-integrated Glassblocks for Active Building Façades. *Int. J. Mod. Eng. Res.* **2015**, *5*, 1–6.
96. Valckenborg, R.M.E.; van der Wall, W.; Folkerts, W.; Hensen, J.L.M.; de Vries, A. Zigzag Structure in Façade Optimizes PV Yield While Aesthetics are Preserved. In Proceedings of the 32nd European Photovoltaic Solar Energy Conference and Exhibition, Munich, Germany, 20–24 June 2016; European Commission: Brussels, Belgium; pp. 647–650.
97. Wall, M.; Probst, M.C.M.; Roecker, C.; Dubois, M.C.; Horvat, M.; Jørgensen, O.B.; Kappel, K. Achieving solar energy in architecture-IEA SHC Task 41. *Energy Procedia* **2012**, *30*, 1250–1260. [CrossRef]

98. Benemann, J.; Chehab, O.; Schaar-Gabriel, E. Building-integrated PV modules. *Sol. Energy Mater. Sol. Cells* **2001**, *67*, 345–354. [CrossRef]

99. Zanetti, I.; Bonomo, P.; Frontini, P.; Saretta, E.; Verberne, G.; Van Den Donker, M.; Sinapis, K.; Folkerts, W. *Building Integrated Photovoltaics. Report 2017*; SUPSI—University of Applied Sciences and Arts of Southern Switzerland, Ed.; SUPSI: Lugano, Switzerland, 2017.

100. Cannavale, A.; Hörantner, M.; Eperon, G.E.; Snaith, H.J.; Fiorito, F.; Ayr, U.; Martellotta, F. Building integration of semitransparent perovskite-based solar cells: Energy performance and visual comfort assessment. *Appl. Energy* **2017**, *194*, 94–107. [CrossRef]

101. Pandey, A.K.; Tyagi, V.V.; Jeyraj, A.; Selvaraj, L.; Rahim, N.A.; Tyagi, S.K. Recent advances in solar photovoltaic systems for emerging trends and advanced applications. *Renew. Sustain. Energy Rev.* **2016**, *53*, 859–884. [CrossRef]

102. Chemisana, D. Building integrated concentrating photovoltaics: A review. *Renew. Sustain. Energy Rev.* **2011**, *15*, 603–611. [CrossRef]

103. Hyder, F.; Sudhakar, K.; Mamat, R. Solar PV tree design: A review. *Renew. Sustain. Energ. Rev.* **2018**, *82*, 1079–1096. [CrossRef]

sustainability

MDPI

Article

The Low-Carbon Transition toward Sustainability of Regional Coal-Dominated Energy Consumption Structure: A Case of Hebei Province in China

Xunmin Ou [1,2], Zhiyi Yuan [1], Tianduo Peng [1,2], Zhenqing Sun [3] and Sheng Zhou [1,*]

[1]   Institute of Energy, Environment and Economy, Tsinghua University, Beijing 100084, China;
     ouxm@mail.tsinghua.edu.cn (X.O.); yuan-zy16@mails.tsinghua.edu.cn (Z.Y.);
     ptd15@mails.tsinghua.edu.cn (T.P.)
[2]   China Automotive Energy Research Center, Tsinghua University, Beijing 100084, China
[3]   School of Economics and Management, Tianjin University of Science and Technology, Tianjin 300222, China;
     sunzq@tust.edu.cn
*   Correspondence: zhshinet@mail.tsinghua.edu.cn; Tel.: +86-10-6279-7626

Received: 19 May 2017; Accepted: 4 July 2017; Published: 6 July 2017

Abstract: CO_2 emission resulted from fossil energy use is threatening human sustainability globally. This study focuses on the low-carbon transition of Hebei's coal-dominated energy system by estimating its total end-use energy consumption, primary energy supply and resultant CO_2 emission up to 2030, by employing an energy demand analysis model based on setting of the economic growth rate, industrial structure, industry/sector energy consumption intensity, energy supply structure, and CO_2 emission factor. It is found that the total primary energy consumption in Hebei will be 471 and 431 million tons of coal equivalent (tce) in 2030 in our two defined scenarios (conventional development scenario and coordinated development scenario), which are 1.40 and 1.28 times of the level in 2015, respectively. The resultant full-chain CO_2 emission will be 1027 and 916 million tons in 2030 in the two scenarios, which are 1.24 and 1.10 times of the level in 2015, respectively. The full-chain CO_2 emission will peak in about 2025. It is found that the coal-dominated situation of energy structure and CO_2 emission increasing trend in Hebei can be changed in the future in the coordinated development scenario, in which Beijing-Tianjin-Hebei area coordinated development strategy will be strengthened. The energy structure of Hebei can be optimised since the proportion of coal in total primary energy consumption can fall from around 80% in 2015 to below 30% in 2030 and the proportions of transferred electricity, natural gas, nuclear energy and renewable energy can increase rapidly. Some specific additional policy instruments are also suggested to support the low-carbon transition of energy system in Hebei under the framework of the coordinated development of Beijing-Tianjin-Hebei area, and with the support from the central government of China.

Keywords: low-carbon transition; regional energy demand; China; Hebei

---

1. Introduction

1.1. General Information of Hebei

Hebei province is in China's north, bordering the Bohai Sea to its east and surrounding Beijing municipality and Tianjin municipality (the area consisting of Beijing, Tianjin and Hebei is called Beijing-Tianjin-Hebei area). In 2015, Hebei ranked 7th among all the nation's 31 provinces in total economic development and 19th in gross domestic product (GDP) per capita (about 6445 US$), which ranks at moderate-to-low level within the whole country. This province has developed a complete industrial system, ranking first in the nation, especially in steel production [1].

Hebei has adopted a typical extensive heavy-chemical industrial development model often seen in China and the quality of the economic development is not high. The industrial structure is in a troubling situation [2], with overwhelming steel industry and a poorly developed tertiary industry. The urbanisation rate of Hebei falls below the national average. The scale of economic development of the major cities in Hebei are smaller than that of Beijing and Tianjin, for instance, even the sum of GDP in 2015 of top three cities (Tangshan, Shijiazhuang and Cangzhou) in Hebei was 1570 billion RMB (Chinese Yuan, the average exchange rate in 2015 is 6.2284 RMB to 1 US$), which is lower than 2270 billion RMB in Beijing and 1653.82 billion RMB in Tianjin in the same year.

Huge differences exist in infrastructure construction and public service levels compared with surrounding Beijing and Tianjin. In particular, a huge gap is formed in the level of economic development in 14 cities (counties) surrounding Beijing and Tianjin, as compared with Beijing and Tianjin themselves [3]. In other words, Hebei is facing gradually urgent pressure on sustainable development in the future [4,5].

1.2. Energy Consumption and CO₂ Emission in Hebei

In 2015, total primary energy consumption in Hebei was 336 million tons of coal equivalent (tce), with a 45% increase from 2006 [6]. Coal consumption maintained a dominant position in primary energy consumption at around 92% from 2006 to 87% in 2015 (Figure 1). In 2015, coal, oil, natural gas, and primary electricity accounted for 86.55%, 7.99%, 3.30%, and 2.16% of total consumption, respectively [6]. Energy consumption per capita in Hebei (3.95 tce) was higher than national average levels (3.13 tce) in 2015 [1,6].

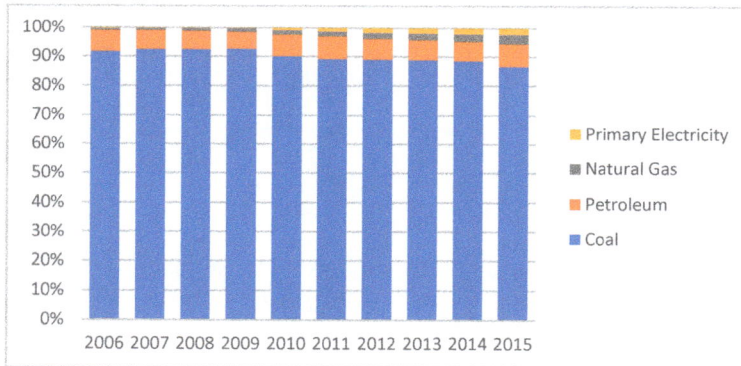

Figure 1. Primary energy consumption composition in Hebei (2006–2015). Data source: [6].

The secondary industry is the foremost contributor to energy consumption in Hebei [7]. Total energy consumption of that industry reached 227 million tce in 2014, accounting for approximately 80% of the total energy consumption. Within the secondary industry sector, energy consumption of six energy-intensive sub-industries (coal mining, oil processing and coking, chemicals, non-metallic, ferrous metal, and electricity and heat) reached 185 million tce in 2015, accounting for approximately 80% of secondary industry sector energy consumption [6]. Energy consumption types in the primary industry include coal, oil (gasoline and diesel), and electricity. In terms of total consumption, energy consumption of primary industry only accounted for 2.1% of total energy consumption in Hebei in 2015 [6]. With regard to energy intensity, energy consumption per unit of GDP of primary industry was 0.19 tce/10,000 RMB in Hebei in 2015 [6]. Energy consumption of the tertiary industry continually increases with the growth of its value-added. Energy consumption for every 10,000 RMB value-added of this industry was 0.24 tce/10,000 RMB [6]. Residential energy consumption mainly involved two sections: energy use in private transport and housing [8–10]. Residential energy consumption per capita in Hebei was 0.46 tce in 2015.

As Figure 2 shows, as the energy consumption increased quickly, CO_2 emission in this area reached about 723 million tons, which is 10 times of the level in 1980 and the average annual growth rate was 6.3% since 2000. Hebei has been under high pressure to control its CO_2 emission increasing trend due to the high ratio of CO_2 emission in Hebei to all of China (about 10%).

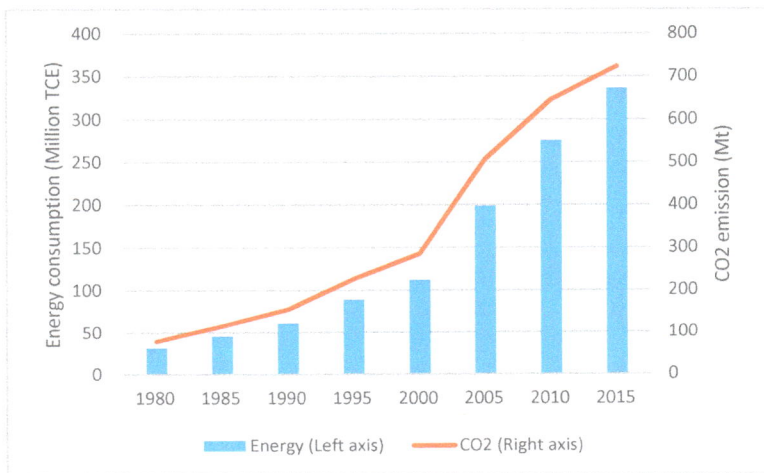

Figure 2. Energy consumption and CO_2 emission in Hebei (1980–2015). Data source: [6,9].

1.3. Requirements of the Beijing-Tianjin-Hebei Area Coordinated Development Strategy

Beijing-Tianjin-Hebei area coordinated development has strategically reached an unprecedented level and the central government of China released the Beijing-Tianjin-Hebei cooperation and development plan in 2015 [11,12]. One of the key drivers is to alleviate the heavy air pollution situation in this area, majorly caused by the large scale of coal use in Hebei, whose SO_2, NOx and dust pollutants accounted 6%, 7% and 10% of the total in China in 2015, respectively [8].

In the future, these two major cities in northern China (Beijing and Tianjin), along with the large industrial province Hebei, will constitute an integrated and developed region [12]. The economic development and resources utilisation in this region will be managed in a coordinated framework. Hebei is responsible for distributing several functions of Beijing, implementing low-carbon energy policies, and accelerating the industrial structure transition [13]. Hebei will also play more significant roles as an ecological service platform as well as an ecological conservation zone [14].

In early 2016, Beijing-Tianjin-Hebei area economic and social development plan during the period from 2016 to 2020 was released and it was the first inter-provincial plan in China based on the "Beijing-Tianjin-Hebei cooperation and development plan" released by the central government of China in 2015. In this plan, the Beijing-Tianjin-Hebei area will be developed in a coordinated way and realise the integrated layout of urban agglomeration development, industrial transformation and upgrading, transportation facilities, and social and livelihood improvement [15].

In April 2017, the central government of China decided to set up Xiong'an New District as a new national-level district. It is planned to develop about 100, 200 and 2000 square kilometre areas in near, medium and long term, respectively, to ease the Beijing non-capital functions, optimise the urban layout and spatial structure of Beijing-Tianjin-Hebei area, and cultivate new driver for innovative development in China [16].

Therefore, Hebei urgently needs to adjust the energy consumption structure, reduce coal use, improve the proportion of clean energy among energy consumption, and achieve diversification of energy supply to decrease fossil energy consumption, CO_2 emission and air pollutants emission [5,17]. The key question is how to fulfil the low-carbon transition in Hebei's energy system from the perspective of the future energy development strategy and policy-decision. Developing the research methodology and modelling tool, and performing in-depth study on the path to the low-carbon transition and corresponding policy suggestions are urgently needed.

1.4. Literature Review of the Studies on Regional Low-Carbon Transition of Energy Structure

Previous research on low-carbon transition of energy system can be mainly categorised into two types: research based on top-down models and research based on bottom-up sector-level analysis.

For the first type, the commonly used optimisation models, including the cost minimisation model, utility maximisation model and computable general equilibrium (CGE) model, are very complex and require abundant data, since cost and economic parameters for various types of energy use technologies are often needed when these models are adopted. For example, International Energy Agency and Nordic Energy Research (2016) [18] focused on energy and carbon technology pathways in the Nordic region through optimisation modelling. Liu et al. (2016) [19] studied the primary energy consumption and carbon emissions in different scenarios at 10-year intervals between 2010 and 2050 in whole China and showed that controlling coal consumption will have an important influence on the control of total carbon emissions and of carbon emission peaking; promotion of non-fossil fuel energies will offer a growing contribution to a low-carbon transition in the medium and long term and the establishment of a low-carbon power system is crucial for the achievement of low-carbon energy transition. Yuan et al. (2014) [20] constructed China's 2050 energy consumption and associated CO_2 reduction scenarios by simulating China's economic development and the consideration on the impacts of urbanisation and income distribution.

For the second type, some studies are conducted by estimating energy consumption and demand through decomposing it sector-by-sector. In these studies, energy efficiency data and structural parameters are essential. For example, Emodi et al. (2017) [21] explored Nigeria's future energy demand, supply and associated GHG emissions from 2010 to 2040 by using the Long-range Energy Alternative Planning (LEAP) model. Linda and Espegren (2017) [22] analysed how various energy and climate measures can transform Oslo into a low-carbon city by employing a technology-rich optimisation model. Zhao et al. (2017) [23] presented a study on the energy system modelling towards renewable energy and low carbon development for Beijing by using the EnergyPLAN tool. Pan et al. (2017) [24] examined China's energy transformation towards the 2 °C global warming goal until 2100 by using integrated-assessment model (the Global Change Assessment Model (GCAM)).

These two common types of methods may not be available for the analysis on sub-region of the developing countries (including China), in which the economic and technological contexts are in a fast-evolution period [25].

1.5. Recent Studies on Hebei's Energy Demand Projection

Most early research specified on Hebei discussed about energy demand by considering overall economy development, and about carbon emission by considering energy structure changing. For example, Du et al. (2015) [7] analysed the dynamic relationship between energy consumption and economic growth by using the time series data of output, capital, employment and energy consumption in Hebei from 1980 to 2012. Niu et al. (2011) [26] analysed the relationship between the carbon emissions and the changes in energy consumption structure in Hebei province, and predicted carbon emissions from 2010 to 2015. Wang and Yang (2015) [14] studied the relationship between regional carbon emissions and industry development of Beijing-Tianjin-Hebei economic band case.

Few studies conduct quantitative analysis on this region's low-carbon transition of energy system since large amounts of data on efficiency and cost are indispensable if optimisation methods or cost minimisation methods are adopted. Because only sectoral activity level data are available at provincial level in China, a useful method for estimating total regional energy consumption can be employed by integrating sectoral estimation of energy consumption and resultant carbon emission with government plans, expert opinions and literature review. For example, there was a study, taken by Institute of Tsinghua University at Hebei (ITUH) [10], aimed to support the strategy decision on the coordinated development of Beijing-Tianjin-Hebei area, involving economic development, energy consumption and air pollution emission. In the study of [10], two scenarios are designed for the future development for Hebei to 2030, based on the different prediction of GDP and its structure. Future energy demand was derived on the method of energy consumption intensity by sector while energy supply structure was determined by assumptions based on trend analysis and central government expectation.

1.6. Scope and Structure of This Paper

Based on the literatures review shown in Sections 1.4 and 1.5, we find it is meaningful to focus on the research for the low carbon transition of energy-system by develop suitable tools for the specific regions such as Hebei. This study will investigate how the more stringent plans can alter the energy structure to affect this regional low-carbon transition using Hebei as a case based on a bottom-up modelling analysis. We focus on the low-carbon transition of Hebei's coal-dominated energy system by estimating its total end-use energy demand, primary energy supply and resultant CO_2 emission up to 2030, by employing an energy demand analysis model based on settings of the economic growth rate, industrial structure, industry/sector energy consumption intensity, energy supply structure, and CO_2 emission factor.

This paper has six sections. Section 2 introduces the methodology, including the model adopted and the calculation equations employed. Section 3 presents the two scenarios defined in this study and their major settings. Section 4 mainly lists the key data and assumptions. Section 5 gives out the results of energy consumption and CO_2 emission, and makes some discussions. The conclusions and policy recommendations are given in Section 6.

2. Methodology

2.1. Framework of Model

The quantitative analysis model based on energy consumption intensity is chosen in the present study, as Figure 3 shows. Future end-use and primary energy demand, and direct and full-chain CO_2 emission by industry/sector are calculated step by step, with the input of GDP, energy intensity of each industry/sector, emission factor of CO_2, conversion and transmission loss rate from primary energy to end-use energy, and energy supply strategy.

That is, for an industry/sector, its future total end-use energy consumption can be estimated by multiplying the energy consumption per unit of GDP or product and GDP output or the products output, respectively. Future total residential end-use energy consumption can be estimated by multiplying the result of residential energy consumption per capita for an urban or rural resident and the corresponding population size.

Based on the end-use energy demand analysis results, primary energy demand and supply can be found with the research on conversion and transmission loss rate from primary energy to end-use energy, and energy supply strategy. Both direct and full-chain CO_2 emission can be calculated with the determination of CO_2 emission factor by energy type.

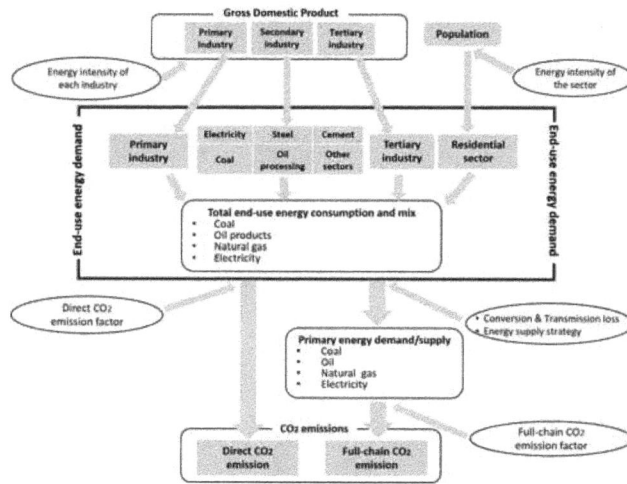

Figure 3. Framework of the quantitative analysis model in this study.

2.2. Calculation Equations

(1) End-use energy consumption

Based on the characteristics of different industries/sector, end-use energy consumption is calculated as following Equations (1)–(6) in this study.

For the year t, the total end-use energy consumption is the sum of end-use energy consumption of four industries/sectors (i) as Equation (1) shows:

$$EndE_{total}^t = \sum_{i=1}^{4} EndE_i^t \tag{1}$$

where $EndE_{total}^t$ represents the total end-use energy consumption, and $EndE_i^t$ is the end-use energy consumption of four industries/sector i in year t. For each industry, it is calculated by multiplying the economic value/goods output by energy intensity.

Equation (2) is adopted for the primary industry:

$$EndE_1^t = GDP_1^t \cdot Intensity_1^t \tag{2}$$

where, for the year t, $EndE_1^t$ is the total end-use energy consumed by primary industry, GDP_1^t is the GDP of primary industry, and $Intensity_1^t$ is the energy consumption per unit of GDP of primary industry.

Equation (3) is adopted for secondary industry:

$$\begin{aligned}
EndE_2^t &= Electricity_{output}^t \cdot Intensity_{Electricity}^t + Steel_{output}^t \cdot Intensity_{Steel}^t \\
&+ Cement_{output}^t \cdot Intensity_{Cement}^t + Coal_{output}^t \cdot Intensity_{Coal}^t \\
&+ GDP_{oil-processing}^t \cdot Intensity_{Coal}^t + GDP_{Chemical}^t \cdot Intensity_{Chemical}^t \\
&+ GDP_{Other-sectors}^t \cdot Intensity_{Other-sectors}^t
\end{aligned} \tag{3}$$

where, for the year t, $EndE_2^t$ is the total end-use energy consumed by secondary industry, $Electricity_{output}^t$ is the electricity output, $Intensity_{Electricity}^t$ is energy consumption per unit electricity output, $Steel_{output}^t$ is the steel output, $Intensity_{Steel}^t$ is energy consumption per unit steel output, $Cement_{output}^t$ is the cement output, $Intensity_{Cement}^t$ is energy consumption per unit cement output,

$Coal^t_{output}$ is the coal output, $Intensity^t_{Coal}$ is energy consumption per unit coal output, $GDP^t_{oil-processing}$ is the GDP of oil processing industry, $Intensity^t_{oil-processing}$ is the energy consumption per unit of GDP of oil processing industry, $GDP^t_{Chemical}$ is the GDP of chemical industry, $Intensity^t_{Chemical}$ is the energy consumption per unit of GDP of chemical industry, $GDP^t_{Other-sectors}$ is the GDP of other sectors in secondary industry, and $Intensity^t_{Other-sectors}$ is the energy consumption per unit of GDP of other sectors in secondary industry.

Equation (4) is adopted for the tertiary industry:

$$EndE^t_3 = GDP^t_3 \cdot Intensity^t_3 \tag{4}$$

where, for the year t, $EndE^t_3$ is the total end-use energy consumed by tertiary industry, GDP^t_3 is the GDP of tertiary industry, and $Intensity^t_3$ is the energy consumption per unit of GDP of tertiary industry.

For residential sector, its end-use energy consumption is calculated by multiplying the population by energy intensity as Equation (5) shows:

$$EndE^t_4 = Pop^t \cdot Intensity^t_4 \tag{5}$$

where Pop^t is the resident population in year t, and $Intensity^t_4$ is the energy consumption intensity per capita in year t.

The end-use energy consumption of each industry/sector can be divided by end-use energy type (j) and the total end-use energy consumption of four industries/sector by end-use energy type can be summed from each industry/sector as Equation (6) shows:

$$EndE^t_{total,j} = \sum_{i=1}^{4} \left(Share^t_{i,j} \cdot EndE^t_i \right) \tag{6}$$

where, for the year t, $EndE^t_{total,j}$ is the total end-use energy type j consumed by the four industries/sectors, and $Share^t_{i,j}$ is the proportion of end-use energy type j in the total end-use energy consumed by industry/sector i.

(2) Primary energy consumption

Based on the calculation results of end-use energy consumption, the total primary energy consumption can be summed from each industry/sector and divided by primary energy type k as Equation (7) shows:

$$PriE_{total} = \sum_{i=1}^{4} \sum_{k=1}^{4} \sum_{j=1}^{4} \left(Con_{j,k} \cdot Share^t_{i,j} \cdot EndE^t_i \right) \tag{7}$$

where $PriE_{total}$ is the total primary energy consumption, and $Con_{j,k}$ is the conversion factor representing the amount of primary energy type k consumed in order to obtain one unit end-use energy type j.

(3) CO_2 emission

Direct and full-chain CO_2 emission can be calculated based on end-use energy consumption and primary energy consumption of each industry/sector, multiplied by different CO_2 emission factors with different energy types, respectively, as Equations (8) and (9) show:

$$CO_2E_{direct} = \sum_{i=1}^{4} \sum_{j=1}^{4} \left(CO_2EF_j \cdot Share^t_{i,j} \cdot EndE^t_i \right) \tag{8}$$

$$CO_2E_{full-chain} = \sum_{i=1}^{4} \sum_{k=1}^{4} \left(CO_2EF_k \cdot \sum_{j=1}^{4} \left(Con_{j,k} \cdot Share^t_{i,j} \cdot EndE^t_i \right) \right) \tag{9}$$

where CO_2E_{direct} is the total direct CO_2 emission, CO_2EF_j is the direct CO_2 emission factor of end-use energy type j, $CO_2E_{full-chain}$ is the total full-chain CO_2 emission, and CO_2EF_k is the full-chain CO_2 emission factor of primary energy type k.

3. Scenario Setting

(1) Economic development settings

Two scenarios have been designed in the study by ITUH [10], in which, the prediction of GDP and its structure in future Hebei without the coordinated development of Beijing-Tianjin-Hebei area, were the key contents. In our study, we will also set two scenarios similar to the ITUH study [10] and agree with not all of its setting but its economic settings in the two scenarios as Table 1 shows. In the conventional scenario, Hebei is assumed to follow the traditional development paradigm for implementing appropriate industrial transformation and upgrading, without considering potential impact to Hebei from Beijing-Tianjin-Hebei area coordinated development strategy: reduction of local air pollutant emission and transferring of high-level industries. Industry on the whole maintains a development trend of seeking progress in stability, while energy-intensive industries also maintain that same trend. In this scenario, the GDP growth rate in Hebei is slightly above the national average in the near term. As the industrial growth rate gradually slows, the GDP growth rate will be roughly comparable with, or slightly lower, than the national average in 2020–2030 [10]. In the coordinated scenario, Hebei vigorously compresses and reduces production capacity in the broad framework of the Beijing-Tianjin-Hebei area coordinated development strategy. It cuts steel production capacity by more than 60 million tons and plans to reduce general steel production capacity to less than 200 million tons by 2020. The province proactively limits secondary industry, particularly development of energy-intensive, highly polluting industries. It assumes industrial transfer from Beijing and Tianjin, and vigorously develops tertiary industry. In this scenario, industrial development in Hebei will be substantially restricted. It is foreseeable that the GDP growth rate in Hebei will drop from above the national average level to the bottom level in the near term. Then, it will gradually return to the national average along with completion of industrial upgrading and vigorous development of tertiary industry [10].

The GDP and other key economic-related data used in two scenarios in future Hebei are shown in Table 2.

Table 1. Scenarios setting on GDP for Hebei development in future.

Scenario	Content
Conventional scenario	GDP growth rate decreases each year, to around 4.5% in 2030.
	Secondary industry still leads the economy and its growth rate will not decrease much; the development rate of tertiary industry will match the GDP growth rate.
	Development of the coal mining and washing, oil processing, coking, and nuclear fuel processing industries are still under strict governmental control, and the output value growth rate is likely zero. Taking into account their inherent factors, the other three industries may have 1–3% output value growth.
Coordinated scenario	GDP growth rate somewhat recovers in a few years, starts to decrease in 2017, and falls to around 3.7% in 2030.
	The secondary industry decline rate is higher than the tertiary industry decline rate. The future tertiary industry growth rate will inevitably exceed the secondary industry growth rate and become the major contributor to the GDP. However, the industrial growth rate will be slightly inferior to the GDP, which is considerably different from in the conventional scenario.
	Energy-intensive industries are further restricted and reduced after 2020. Zero growth of six such industries is gradually achieved during 2020–2030. The GDP in Hebei no longer relies on these industries as a driver for development and the growth rate gradually becomes zero.

Data source: [10].

(2) Energy demand and supply related settings

The energy intensity data in future by industry/sector of each year will not be set differently varied from scenario. As a result, end-use energy demand in future by industry/sector of each year will be different in the two scenarios if their economic and production output of each year are different in different scenario.

In the conventional scenario, the future energy supply strategy will follow the local development plan before 2020, and will keep the coal supply stable from 2020 to 2030 while expand clean energy (natural gas, renewable energy, nuclear power and transferred electricity) to fulfil the incremental energy demand during this period.

In the coordinated scenario, the future coal supply will decrease at a quicker speed than the local development plan before 2020 and will continually decrease by one half from 2015 to 2030. Natural gas, nuclear power and transferred electricity in this scenario will be expanded to the bigger scale than the level in the conventional scenario, while the renewable energy will expand to a bit smaller scale than the level in the conventional scenario, though the percentages of renewable energy in total energy supply of each year will be the same in the two scenarios.

Table 2. GDP and other key economic data in two scenarios in future Hebei.

Item	Unit	Conventional Scenario				Coordinated Scenario			
		2015	2020	2025	2030	2015	2020	2025	2030
GDP	Trillion RMB (constant price in 2015)	2981	4432	6010	7732	2981	4200	5404	6610
Share of GDP									
Primary industry	%	10	8	7	6	10	9	8	7
Secondary industry	%	54	59	62	65	54	53	52	50
Among which, the energy-intensive sub-industries	%	19	18	15	12	19	16	13	11
Tertiary industry	%	36	33	31	29	36	38	40	43
Output of Key Products/Sub-Industries									
Electricity production	Trillion kWh	233	284	378	473	233	254	480	571
Steel production	Million ton	142	181	181	181	142	147	126	108
Cement production	Million ton	124	129	129	129	124	112	110	91
Coal production	Million ton	97	118	118	118	97	110	105	100
Oil processing industry output	Trillion RMB (constant price in 2015)	67.3	94.6	128.3	165.1	67.3	89.7	115.4	141.1
Chemical industry output	Trillion RMB (constant price in 2015)	45.8	64.4	87.3	112.3	45.8	61.0	78.5	96.0

Data source: [10].

4. Key Data and Assumption

4.1. Macro-Economic, Demographic and Urbanisation Rate Assumption

Economic output is referred from the ITUH study [10] on future development of Hebei province, as Tables 1 and 2 show, in the two designed scenarios as mentioned in the Section 3. Total population and urbanisation rate data are referred from the yearbook and future plans of Hebei [6,27] as Table 3 shows. Population in Hebei will sustainably increase and peak after 2030, since China began to relax the family planning (one-child) policy in 2014. Meanwhile, urbanisation rate in Hebei will increase rapidly, since China has determined to relax the conditions for transference from rural to urban household registration, and employment opportunities in cities are much more abundant than that in rural areas.

Table 3. Population and urbanisation rate of China from 2010 to 2030 in Hebei.

	Unit	2015	2020	2025	2030
Population	million	74.45	77.14	79.93	82.82
Urbanisation rate	%	52	55	61	66

Data source: [6,27].

4.2. Energy Intensity Data

The energy intensities of different industries/sectors in the future are calculated by authors by referring [8,28–33] (see Table 4). The details are shown in the following paragraphs.

Table 4. Energy intensity of different industry/sector in Hebei (2015–2030).

	Unit	2015	2020	2025	2030
Primary Industry					
Energy consumption per unit of GDP	tce/10,000 RMB (constant price in 2015)	0.19	0.17	0.15	0.14
Secondary Industry					
Energy consumption per unit of product: electricity	gce/kWh	320	310	300	290
Energy consumption per unit of product: steel	tce/ton	0.54	0.39	0.34	0.32
Energy consumption per unit of product: cement	kgce/ton	92	80	74	70
Energy consumption per unit of product: coal mining and processing	kgce/ton	8.5	7.5	6.6	6.0
energy consumption per unit of value-added: oil processing industry	tce/10,000 RMB (constant price in 2015)	0.83	0.78	0.78	0.78
Energy consumption per unit of value-added: chemical industry	tce/10,000 RMB (constant price in 2015)	1.38	1.18	1.16	1.16
Energy consumption per unit of value-added: other industry	tce/10,000 RMB (constant price in 2015)	1.47	0.99	0.78	0.63
Tertiary Industry					
Energy consumption per unit of value-added	tce/10,000 RMB (constant price in 2015)	0.24	0.20	0.16	0.14
Residential Sector					
Energy consumption per capita	tce/person	0.46	0.82	1.25	1.43

Note: (1) The data in 2015 are derived from the energy consumption data in [6,8] and economic/production output data in [6,9]; and (2) the data beyond 2015 are calculated by authors based on the trend analysis by referring to [29–33] with details explained in main text.

(1) Primary industry

Energy consumption per unit of GDP of primary industry was 0.19 tce/10,000 RMB in Hebei in 2015 [6,8,9]. The electricity consumption intensity of primary industry in Hebei exceeded the national average, while the corresponding oil consumption intensity was below the national average [28]. With the saturation of mechanisation in the primary industry, process of agricultural modernisation will mainly depend on development of biotechnologies and information technologies in future. Consequently, energy consumption per unit of GDP of primary industry will potentially further decrease in the future. Energy consumption per unit of GDP of primary industry is set to 0.14 tce/10,000 RMB for 2030, which declines by nearly 26% compared with that in 2015 (0.19) and approaches the national level in 2014 (0.13) [9].

(2) Secondary industry

Substantial decreases (about 10–40%) will occur in energy consumption per unit of product in the electricity, steel, cement, and coal mining and processing industries in the period 2015–2030 [29–31]. Future downswing of energy consumption per unit of product is mainly derived from technology progress. Current energy consumption per unit of product in the electricity industry in Hebei is 20% higher than that in the advanced world level. Adopting advanced technology will help improve energy efficiency in Hebei. In the steel industry, there is an enormous potential of energy consumption reduction if the proportion of short-flow production line, which uses scrap steel as feedstock, increases and energy-saving technologies are adopted. Hebei will spread several advanced energy-saving technologies in the cement industry, for instance, new-dry-process, and gradually bridge the gap between energy consumption level in Hebei and the advanced world level. As for the coal mining and processing industry, Hebei will continuously improve the mining level, enlarge the application of energy-saving technologies, and gradually improve the energy consumption level.

Energy consumption per unit of value-added in oil processing and chemical industry dropped considerably, from 6.1 and 7.2 tce/10,000 RMB in 2005 to 0.83 and 1.38 tce/10,000 RMB in 2015, respectively. Given that the decline in energy consumption intensity is not unlimited, the rate of decline tended to slow from 2015 to 2030. Energy consumption intensity in other sub-industries will be lower than that in chemical industry in future [10,31]. Future downswing of energy consumption per unit of value-added in these key sub-industries is mainly derived from technology progress and proportion increase of high-value products, for instance, a greater proportion of high value-added petrochemicals will be produced in oil processing and chemical industry.

(3) Tertiary industry

Energy consumption intensity of tertiary industry in Hebei was 0.24 tce/10,000 RMB in 2015. Oil product consumption was 0.057 tons/10,000 RMB, natural gas consumption was 11.68 m^3/10,000 RMB, and electricity consumption was 313 kWh/10,000 RMB [6,8,9]. The future energy consumption intensity of the tertiary industry in Hebei shows a downward trend [32,33]. Based on the current level of energy consumption intensity of the tertiary industry in the country and at the international advanced level [33], we assume that energy consumption intensity of tertiary industry in Hebei will drop by nearly 40% in 2030 compared with 2015.

Developing tertiary industry significantly orients optimisation of industrial structure and transformation of major engine of economic growth in Hebei. As the process of industrial upgrading accelerates, high value-added sub-industries will develop rapidly, proportion of coal use among total energy consumption will decrease simultaneously, and proportion of electricity and steam demand among total end-use energy consumption will increase steadily in tertiary industry. As the added value of transportation industry will increase rapidly, the demand of oil products and other fuels will increase consequently.

(4) Residential sector

In 2015, Hebei has expanded to a total population of about 74.45 million and reached a 52% urbanisation level, with a total urban population of around 38 million [9]. Residential energy consumption per capita in Hebei was 0.46 tce, with a combination of 0.50 tce per capita for urban resident and 0.40 for rural resident [8]. Overall, Residential energy consumption per capita in Hebei will be 1.43 tce in 2030, growing by approximately 200% compared with the level in 2015 [10].

Increase of residential energy consumption per capita results from upturn living standards, urbanisation rate increase and development of electrification level. Residential demand of electricity and natural gas will increase, while demand of coal and liquefied petroleum gas will decrease. Along with increased vehicle population and frequent commuting, oil product consumption will also increase rapidly in future in Hebei.

4.3. Other Key Data and Assumptions

(1)   End-use energy consumption structure of secondary industry

End-use energy consumption structure of secondary industry shown in Figure 4 is adopted for energy-intensive sectors in Hebei in 2015 [8]. In the future, this kind of structure will be optimised by increasing the share of clean energy (electricity and natural gas).

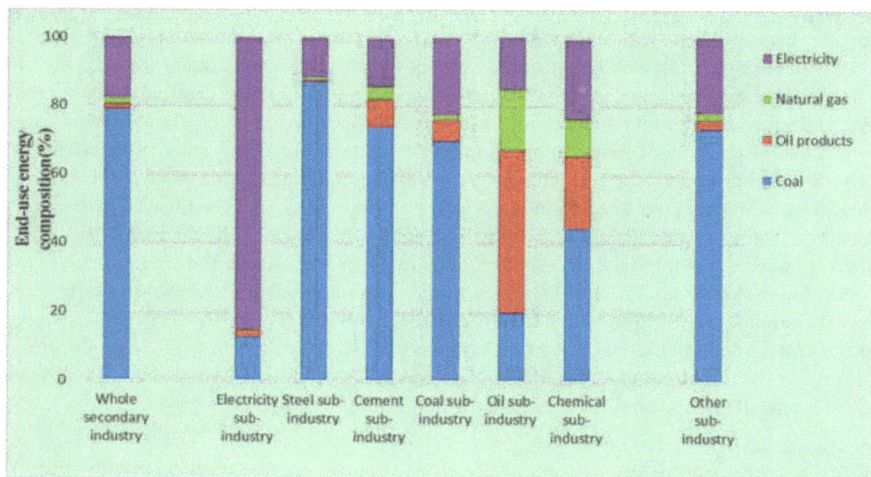

Figure 4. End-use energy consumption structure of secondary industry and its sub-industries in 2015. Data source: [8].

(2)   Status and development plan of non-fossil energy in Hebei

In general, renewable energy resources are abundant in Hebei.   There are ample wind energy resources in Hebei, in which technically exploitable resource surpasses 80 million kilowatts. Solar energy resources in Hebei rank in the forefront in China, with 90 million kilowatts available for exploitation. Biomass resources are abundant in Hebei. Total amount for energy-oriented use is over 20 million tons, within which crop straw and forestry resources are 10.46 million tons and 5.7 million tons, respectively. Annual surficial geothermal resource in Hebei is equivalent to 285 million tce, with 11 million tce available for exploitation. Geothermal resource reserve in middle-deep beds is equivalent to 23.52 billion tce, with 4.97 billion tce available for exploitation. Potential exploitation amount of capacity of pumped storage electricity plant is over 16 million kilowatts [34,35].

Renewable energy resource in Hebei has been well developed and utilised [35,36]. The gross of renewable energy use in Hebei increased from 4 million tce in 2010 to approximately 10 million tce in 2015, while proportion that renewable energy consumption accounts for of primary energy consumption rose from 2.4% to about 5% during the same period. In 2015, generated electricity from renewable energy in Hebei was 21.3 billion kilowatts, accounting for 6.7% of total electricity consumption.

Hebei will exert efforts to raise renewable energy supply in near future according to development plan [34]. In 2020, Utilisation amount of renewable energy will reach 23 million tce, and non-fossil energy consumption will account for 7% of total energy consumption. Generated electricity from renewable energy will account for over 13% of total electricity consumption in Hebei.

Hebei has also exerted efforts to put forward low-carbon transition by constructing several nuclear plants [36]. Site selection of nuclear power plants in Kuancheng, Funing, Qianxi and Haixing has already been accomplished. Obvious progress has been made in early stage works of Haixing nuclear plants' construction and the first unit will be put into production in December 2020, while the second one will be commissioned 10 months later. Four other nuclear plants may also be constructed, and the total installed capacity will be 7.5 million kilowatts.

As mentioned in the Section 3 (Energy demand and supply related settings of future development scenario setting), in the conventional scenario, the future energy supply strategy in this study will follow the renewable energy development plan before 2020 and expand renewable energy and nuclear power, along with transferred electricity, to keep the coal supply stably from 2020 to 2030.

In the coordinated scenario, because the future coal supply will be decreased at a quicker speed than the local development plan before 2020 and will continually decrease to cut one half from 2015 to 2030, nuclear power, along with transferred electricity, in this scenario will be expanded to the bigger scale than the level in the conventional scenario, while the renewable energy will expand to a bit smaller scale than the level in the conventional scenario, though the percentages of renewable energy in total energy supply of each year will be same in the two scenarios.

From the methodological point of view, we will build the energy development plans summarised here to our primary energy supply projections up to 2030 by following approach. For a specific year during 2016 to 2030, we will analyse the primary energy supply by type one by one, based on the result of total demand:

- Coal supply both for direct use and power generation;
- Oil supply;
- Natural gas supply for direct use and power generation; and
- Non-fossil energy demand for power generation and other energy service.

During the period from 2015 to 2020, energy supply in the conventional scenario basically conform to the relevant development plan for 2015–2020 promulgated by Hebei government. In the coordinated scenario, coal assumption will reduce by 20%, and energy consumption of natural gas, nuclear power and electricity transferred from other provinces will consequently increase to ensure total energy consumption in Hebei in the same period. Besides, we assume the proportion of renewable energy among total energy consumption in the coordinated scenario remains the same as that in the conventional scenario, while their actual quantity in the coordinated scenario will slightly decrease compared to that in the conventional scenario.

During the period from 2020 to 2030, the major variable considered in energy supply is coal assumption. In the conventional scenario, coal demand will remain the same from 2020 to 2030, while oil product demand will increase slightly. Demand of natural gas will employ an annual 5% increase. Electricity and steam will be guaranteed by rapid development of nuclear power and moderate development of renewable energy. Electricity transferred from other provinces will also help meet the redundant electricity demand. In the coordinated scenario, coal demand from 2020 to 2030 will reduce by about 50% compared with that in 2015. Oil product demand will slightly increase. Natural gas demand will double its original status in 2015. Electricity demand will be guaranteed by rapid development of nuclear power and moderate development of renewable energy. Electricity transferred from other provinces will also help meet the redundant electricity demand.

To sum up, proportions of renewable energy demand among total energy demand in both the conventional scenario and coordinated scenario are about 5%. Nuclear power will gradually develop and finally account for about 50% of total available exploited resources in 2030 in the conventional scenario, while this proportion in the coordinated scenario will be about 67%.

(3) Loss rates of coal and oil during production and transmission

It is supposed that loss rates of coal and oil during production and transmission are 9% and 3%, respectively [9]. The primary energy demand can be derived by these loss rates and the obtained end-use energy demand by Ref. [25].

5. Results and Discussion

5.1. Results of End-Use Energy Consumption

(1)   Total end-use energy demand

Tables A1 and A2 and Figure 5 indicate end-use energy consumption by industry/sector and fuel type in the conventional and coordinated scenarios.

In the conventional scenario, the total end-use energy consumption in Hebei will increase from 228 million tce in 2015 to 272 million tce in 2030, and the growth ratio will be 19%. The ratio of energy consumption of secondary industry will drop from 76% to around 66% in 2030. Coal will be gradually replaced by gas and electricity. Therefore, its proportion among the total energy consumption will decrease from approximately 65% in 2015 to 45% in 2030.

In the coordinated scenario, the total energy consumption in Hebei in 2030 can be 235 million tce (about 37 million tce less than that in the conventional scenario). Energy structure can be sufficiently optimised, since the proportion of secondary industry sector's energy consumption among the total energy consumption can drop to below 60%, and the proportion of coal among the total energy consumption can decrease to 32% in 2030, which are much better than the results in the conventional scenario.

Coal utilisation of end-use energy consumption in the coordinated scenario is 40% below that in the conventional scenario, while natural gas utilisation in the coordinated scenario is 23% more than that in the conventional scenario, resulting from gradually incremental utilisation of natural gas substituting the direct use of coal.

Oil utilisation of end-use energy consumption in the coordinated scenario is 12% more than that in the conventional scenario, while electricity utilisation in the coordinated scenario is 2% more than that in the conventional scenario, resulting from growth of the tertiary industry.

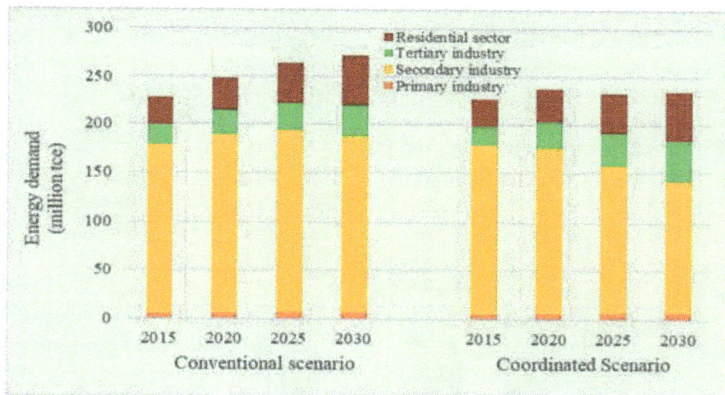

(a) By industry/sector

Figure 5. *Cont.*

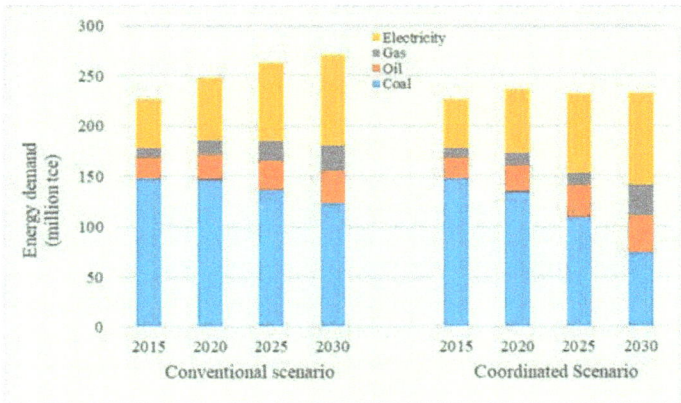

(**b**) By type

Figure 5. End-use energy demand in Hebei from 2015 to 2030 in two scenarios by industry/sector (**a**) and by type (**b**).

(2)   Primary industry

It can be estimated that future energy demand in primary industry of Hebei will remain stable, from about 6 million tce in 2015 to 7 million tce in 2030.

(3)   Secondary industry

In the conventional scenario, energy consumption of the secondary industry in Hebei will gradually increase, from 172 million tce in 2015 to 183 million tce in 2020, and then 181 million tce in 2030 as Figure 6 shows. Energy structure of the secondary industry in Hebei is based on coal and electricity, at the proportions of 75% and 20%, respectively, in 2015; the proportions of oil and gas were lower, at 2% and 3%, respectively. Owing to the increase in non-energy-intensive industry energy consumption, particularly the substantial increase in the proportions of oil and gas in non-energy-intensive sub-industries, the proportion of gas in the electricity sub-industry will increase markedly; the proportions of coal, electricity, oil, and gas in 2030 will be 56%, 33%, 3%, and 8%, respectively.

In the coordinated scenario, total energy consumption of the secondary industry sector will decrease, reaching 170 million tce in 2020 and 135 million tce in 2030. In 2030, proportions of coal, electricity, oil, and gas will be 38%, 44%, 4%, and 14%, respectively. In the coordinated scenario, the proportion of energy consumption will decrease by 39% for coal and increase by 24% for electricity; the increases for oil and gas will be 2% and 12%, respectively, in 2030 compared with 2015.

Energy consumption of energy-intensive industries accounted for 80% of total energy consumption of the secondary industry sector in 2015. In the conventional scenario, the proportion will drop to about 60% and about 50% in 2020 and 2030, respectively; in the coordinated scenario, to about 55% and 40% in 2020 and 2030, respectively.

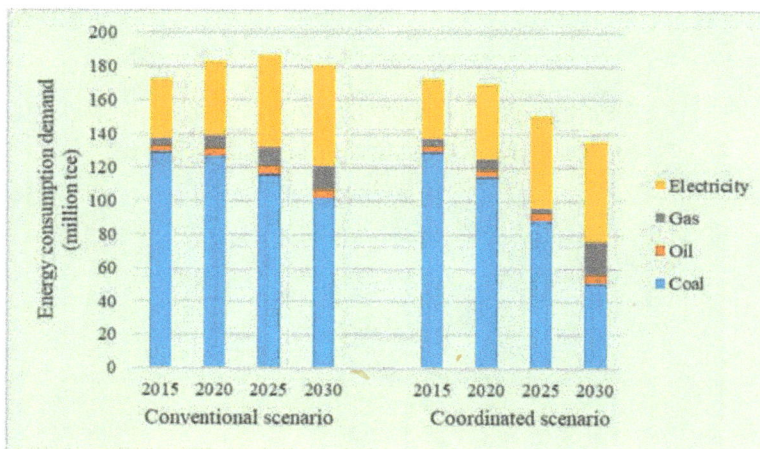

Figure 6. End-use energy demand of Hebei secondary industry in 2015, 2020, 2025 and 2030 in the two scenarios.

(4)   Tertiary industry

End-use energy demand of the tertiary industry in Hebei is estimated to grow from 21 million tce in 2015 to around 33–41 million tce in 2030; a growth rate of around 56–106%. This study shows that energy demand in tertiary industry will rapidly increase, and its proportion in total energy demand will increase from 9% in 2015 to around 12–18% in 2030.

(5)   Residential sector

Future energy demand of Hebei residents in Hebei is estimated to grow from 29 million tce in 2015 to 51 million tce in 2030 with a growth rate of around 75%. This study shows that future energy consumption of Hebei residents will also increase rapidly, with increasing demand for electricity, oil and gas, among the different types of high-quality energy.

5.2. Results of Primary Energy Demand

Table A3 and Figure 7 indicate primary energy consumption by industry and fuel type in the conventional and coordinated scenarios. Primary energy consumption will increase from 336 million tce in 2015 to 471 million tce in 2030 in the conventional scenario. In the coordinated scenario, primary energy consumption will be 431 million tce in 2030. The growth ratio in both scenarios will be 40% and 28%, respectively. Primary energy consumption of the secondary industry will still dominate as the major contributor to the total primary energy consumption, and it will account for 86% and 84% of the total primary energy consumption in 2030 in the conventional scenario and coordinated scenario, respectively, while the proportion in 2015 was 87%.

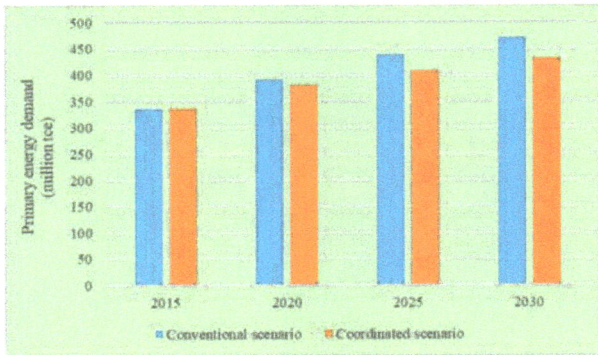

Figure 7. Primary energy demand in Hebei (2015–2030).

5.3. Results of Primary Energy Supply Structure in Future

Primary energy supply in future will be more diversified and cleaner in the defined two scenarios as shown in Table A3 and Figure 8.

Coal consumption used to be 314 million tons (equivalent to 260 million tce) in 2015, and will decrease to 150 million tons (equivalent to 125 million tce) in 2030 in the coordinated scenario, merely accounting for about 30% of total energy supply.

Utilisation amount of natural gas will increase from 8.5 billion cubic meters in 2015 to 40–65 billion cubic meters in 2030, and account for 10–20% of total energy consumption.

Utilisation amount of electricity transferred from other provinces will increase rapidly, and its proportion among total electricity consumption will increase from 28% in 2015 to over 30% in 2030. The quantity will increase from 90 trillion kWh in 2015 to 200–300 trillion kWh.

As for nuclear power, several nuclear plants will be put into production from 2020, and the proportion of nuclear power among total electricity supply will be 14% and 28% in 2020 and 2030, respectively.

Energy supply from renewable energy will develop smoothly, increasing from 10 million tce in 2015 to about 20 million tce in 2030. As Table A4 indicates, in the conventional scenario, non-fossil energy supply ratios in Hebei are 7% and 13% in 2020 and 2030, respectively. In contrast with the conventional scenario, the non-fossil energy supply ratios in the coordinated scenario fulfil the goal set by Chinese government and reach the average level in the country.

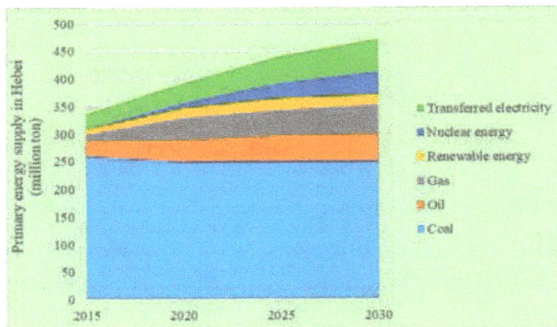

(**a**) in the conventional scenario

Figure 8. *Cont.*

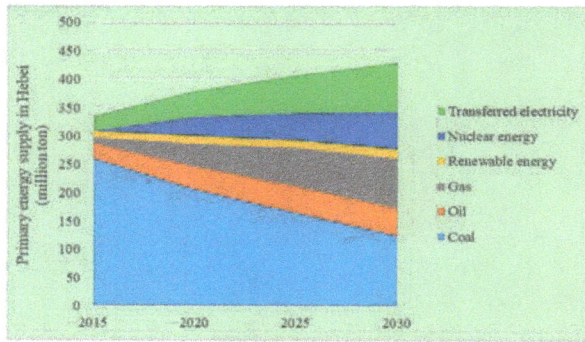

(b) in the coordinated scenario

Figure 8. Primary energy supply in Hebei (2015–2030) in the conventional scenario (**a**) and in the coordinated scenario (**b**).

5.4. Results of CO_2 Emission

(1) Direct CO_2 emission

As Figure 9 indicates, in the conventional scenario, direct CO_2 emission will increase slightly from 444 million tons in 2015 to 456 million tons in 2020, and then peak, decreasing to 432 million tons in 2030. In the coordinated scenario, direct CO_2 emission will constantly decrease from 444 million tons to 323 million tons in 2030, the rate of descent reaching 27%.

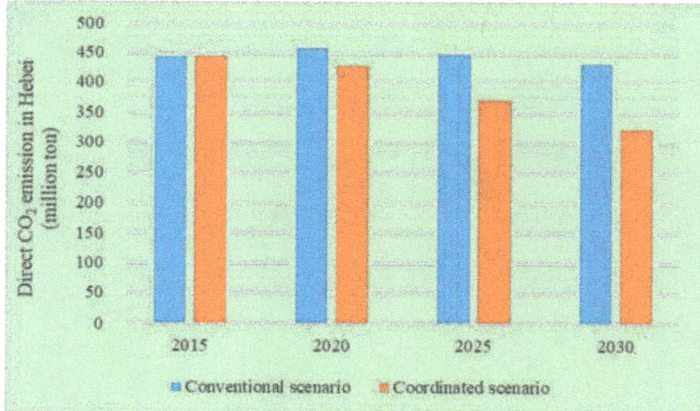

Figure 9. Direct CO_2 emission in Hebei (2015–2030).

(2) Full-chain CO_2 emission

In the conventional scenario, full-chain CO_2 emission will increase from 831 million tons in 2015 to 1027 million tons in 2030, and the growth ratio will reach 24%, while, in the coordinated scenario, full-chain CO_2 emission peaked in 2025, and will constantly decrease to 916 million tons in 2030. Detailed results are shown in Figure 10.

It should be noted that CO_2 emission of electricity transferred from other provinces is considered here. The CO_2 emission factor of electricity adopted here is that in the north China electricity grid, which Hebei electricity grid is affiliated with [37]. CO_2 emission of electricity transferred from other provinces account for about 10% of total full-chain CO_2 emission in Hebei.

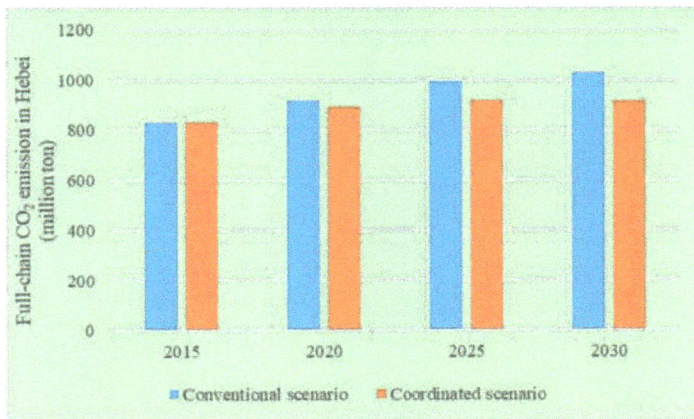

Figure 10. Full-chain CO_2 emission in Hebei (2015–2030).

(3) Accounted CO_2 emission

The accounted CO_2 emission in Hebei in 2015–2030 is shown in Table A4. In the conventional scenario, CO_2 emission will increase from 723 million tons in 2015 to about 800 million tons in 2030, and the growth ratio will reach 10%, while, in the coordinated scenario, CO_2 emission will constantly decrease to 528 million tons in 2030.

Accounted CO_2 emission, which is reported to the central government of China, can be calculated following the standard process [38] by counting up each type of end-use energy consumption multiplied by homologous CO_2 emission factor. The projection of electricity CO_2 emission factor is important during the accounted CO_2 emission calculation. Electricity CO_2 emission factors are different in different regions' power grids in China. The emission factors of electricity used in Hebei in 2015, 2020 and 2030 are 0.837, 0.755 and 0.725 kg/kWh, respectively, same as North China Power Grid, which Hebei is attached to [38].

5.5. CO$_2$ Reduction Contribution Analysis

It should be noted that clean trend of energy structure can help reduce CO_2 emission resulted from energy consumption, especially when nuclear power and electricity generated from renewable energy are promoted. CO_2 emission in the coordinated scenario will keep decreasing, since nuclear power and electricity generated from other renewable energies will bring zero CO_2 emission.

Supposed that CO_2 emission from electricity in Hebei is calculated based on its actual production status, CO_2 emissions from 2015 to 2030 in Hebei are shown in Figure 11. Comparing CO_2 emissions in the coordinated scenario with that in the conventional scenario, the reduction ratios will be 14%, 23% and 34% in 2020, 2025 and 2030, respectively.

GDP changes will maintain about 1/3 of contribution to CO_2 emission reduction in the long run. Increasing transferred electricity's contribution to CO_2 emission reduction will gradually increase, and turn out to be about 1/4 in 2030. In the short term, non-fossil energy development will be the most effective method to reduce CO_2 emission, since non-fossil energy development can sharply reduce the demand of coal electricity, and it will account for about 40% and about 70% of CO_2 emission reduction in 2020 and 2030, respectively.

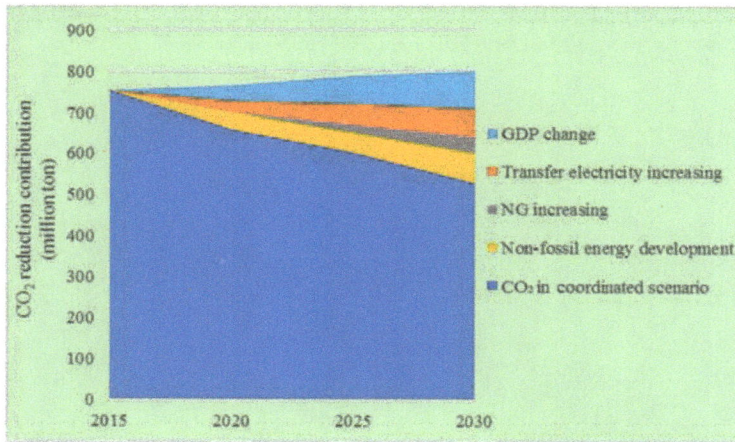

Figure 11. CO_2 reduction contribution of different factors (2015–2030).

5.6. Comparison of Energy Demand Results with Other Studies

As Figure 12 shows, the results of end-use energy demand for 2020 and 2030 in the two defined scenarios of this study are larger than the results in the study of [10]. The main reason for the difference is that a more detailed analysis of the seven sub-sectors of secondary industry has been carried out in this study. The declining trend of energy intensity coefficient of secondary industry used in this study is more moderate than the study of [10]: the decrease rate from 2015 to 2030 is about 50% in this study while about 70% in the study of [10].

The end-use energy demand for 2020 and 2030 in Hebei reported in Zhang Xu [38], which studied the whole China's energy demand up to 2030 by considering interactions among different provinces, is on the result-range of the two defined scenarios of both this study and the study [10]. More efforts on controlling the energy demand in Hebei have been required to accelerate the coordinated development for Hebei in the coordinated scenario in this study than the study of Zhang Xu [38], which has partly considered controlling the energy demand in Hebei to help the regional synergistic development.

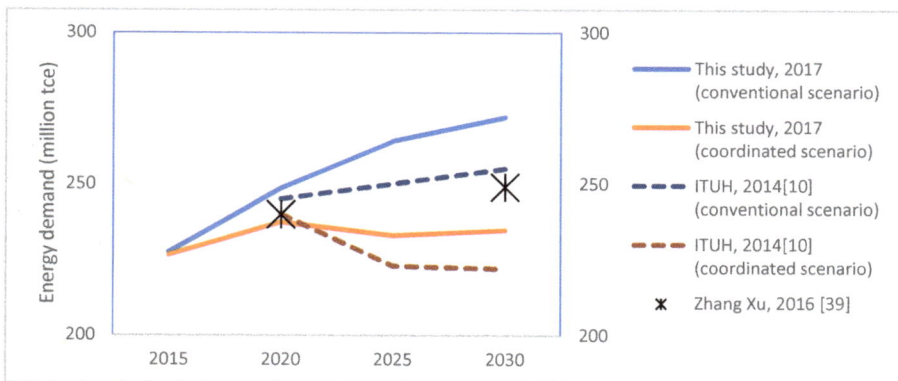

Figure 12. End-use demand in Hebei research results in different studies.

5.7. Comparison of CO₂ Emission and Energy Structure in Hebei and Whole China in Future

As Table A4 shows, Hebei's CO_2 emission (accounted number following the inventory guideline) has peaked in the year of 2015 and will decrease about 30% from 2015 to 2030 in the coordinated scenario, accompanied by lower proportion of coal and higher of non-fossil energy in the energy structure than in in conventional scenario. The proportion of coal (which is carbon intensive) in the total primary energy consumption will fall from around 80% in 2015 to 53% in 2030 and the proportion of non-fossil energy (which is carbon free) will finally grow to 13% in 2030 in the conventional scenario. In the coordinated scenario, the proportion of coal in the total primary energy consumption will fall to about 30% in 2030 and the proportion of non-fossil energy will finally grow to 20% in 2030.

As mentioned in Section 5.4, according to the calculation method proposed by provincial accounted CO_2 emission reports to the central government of China, indirect CO_2 emissions from transferred electricity is equal to the product of actual quantities of transferred electricity times average emission factor of its regional power grid.

China government has pledged to peak the CO_2 emission around 2030 and planned to enforce the non-fossil energy ratio of total energy consumption to reach 12% and 20% in 2020 and 2030, respectively, while decreasing coal consumption to decrease its proportion in the total primary energy consumption below 50% [39,40].

It is found that the effect of energy saving and GHG reduction in future in Hebei can surpass the national average level through efforts to decrease the total energy consumption of Hebei, increase the clean energy and greatly optimise the energy structure in the coordinated development scenario.

5.8. Possibility of Realiazation of Coordinated Scenario in Hebei

Whether the coordinated scenario in Hebei can be realised or not mainly depend on how much effort Hebei, the whole area of Beijing-Tianjin-Hebei, and even whole China exert, since this scenario is quite ideal. Major risks that can jeopardize the coordinated scenario's results include: optimisation level of the industrial structure, actual natural gas supply and infrastructure construction level that can be ensured, development rate of local nuclear power plants, transferred electricity that can be guaranteed and Hebei's bearing capacity for future increase of electricity price.

Some critical factors that help evade the risks mentioned above include: coordinated development of the whole Beijing-Tianjin-Hebei area, introduction of high-level industries and strong energy supply and financial support from the central government of China.

5.9. Major Implication of the Findings on Hebei Case for Other Regions

The findings in this study of Hebei can be applied to understand the energy system low-carbon transition in the regions with similar economy and energy structure in China and beyond. A deeply collaborated development mode will bring out CO_2 reduction benefit with stronger energy consumption constraints. There are three key factors to driven the low-carbon transition of the regional energy system dominated by coal: GDP structure optimisation, the introduction of external low-carbon energy and the improvement of local low-carbon energy development and utilisation.

In addition, we should pay more attention to the co-benefit of environmental protection with low-carbon energy system transition: the substitution of non-fossil energy on fossil energy can reduce SO_2, NOx and dust soot emissions.

6. Concluding Remarks

This paper adopts a provincial energy demand analysis model based on energy intensity analysis, conducts a quantitative analysis on energy consumption and discusses the low-carbon transition path of Hebei province in China. The research method can by referred by other studies to research on similar energy demand of a specific region without abundant sectoral data.

It is found that future energy consumption and CO_2 emission in Hebei are substantially different when following different development paths. Several achievements can be realised though efforts. The total primary energy consumption and accounted CO_2 emission will be 431 million tce and 528 million ton in the coordinated scenario in 2030, 40 million tce and 272 million ton less than in the conventional scenario. The energy structure of Hebei will be optimised, since the proportion of coal in total primary energy consumption can fall from around 80% in 2015 to 53% and 29% in 2030 in the two defined scenarios and the proportions of transferred electricity, natural gas, nuclear energy and renewable energy can increase rapidly.

In the coordinated scenario, Hebei's CO_2 emission has peaked in 2015 and will decrease about 30% from 2015 to 2030 accompanied by lower proportion of coal and higher of non-fossil energy in the energy structure than in in conventional scenario.

Hebei is supposed to fulfil the objective toward sustainable development and obtain the effect of energy saving and GHG reduction which can surpass the national average level through efforts to decrease the total energy consumption of Hebei, increase the clean energy and greatly optimise the energy structure in future.

There are three key factors driving the low-carbon transition of the regional energy system dominated by coal: GDP structure optimisation, the introduction of external low-carbon energy and the improvement of local low-carbon energy development and utilisation.

Some specific additional policy instruments are suggested to support the low-carbon transition of energy system in Hebei under the framework of the coordinated development of Beijing-Tianjin-Hebei area:

(1) Accelerate the backward production capacity elimination and compression of the scale of energy-intensive industries currently by taking over high-end industry from Beijing and Tianjin, ensuring the steady development of the economy.

(2) Transfer more electricity from outside than planned, by determining the source of electricity supply, constructing the electricity transmission channel and developing the low-price mechanism.

(3) Exert efforts to utilise natural gas at a bigger scale than planned by determining natural gas supply source, building the piping with the full support of the central and local government and implementing the low-price mechanism to promote the efficient supply of natural gas and large-scale use for electricity generation, industry and residents.

(4) Introduce nuclear power rapidly, by constructing several nuclear power parks in the coastal and inland area gradually and operating the nuclear power plant safely and efficiently in near future.

It is worth noting that several listed instruments cannot be implemented by only Beijing-Tianjin-Hebei area, and require overall coordination of the central government of China. These instruments include: assurance of natural gas supply source, support for quick nuclear power development in Hebei and guarantee of transferred electricity source to Hebei.

In this way, revolutionary change in the energy system can support the coordinated targets of economic growth and CO_2 emission control, and environmental governance by 2030, and benefit the sustainable development of this region dominated by coal energy currently.

Acknowledgments: This project was co-sponsored by key project of the China National Social Science Foundation (16AGL002) and the National Natural Science Foundation of China (71690240, 71373142 and 71673165).

Author Contributions: Xunmin Ou and Sheng Zhou conceived and designed the research framework; Xunmin Ou performed the model development and analyzed the data; Xunmin Ou, Zhiyi Yuan, Tianduo Peng, Zhenqing Sun and Sheng Zhou wrote the paper.

Conflicts of Interest: The authors declare no conflict of interest.

Appendix A

Table A1. End-use energy consumption demand up to 2030 in conventional scenario (million tce).

	2015	2020	2025	2030
Primary Industry				
Sub-total	5.59	6.26	6.75	6.99
Coal	1.31	1.80	2.35	3.01
Oil	2.87	3.03	3.04	2.81
Gas	0.00	0.00	0.00	0.00
Electricity	1.20	1.18	1.07	0.81
Secondary Industry				
Sub-total	172.25	183.24	186.68	180.53
Coal	129.46	127.11	115.65	101.51
Oil	3.52	4.34	5.14	5.36
Gas	4.33	7.66	11.35	13.97
Electricity	34.95	44.13	54.54	59.70
Tertiary Industry				
Sub-total	21.06	25.44	29.10	33.24
Coal	5.21	6.29	7.20	8.23
Oil	10.61	12.81	14.66	16.75
Gas	0.98	1.18	1.35	1.54
Electricity	3.47	4.20	4.80	5.48
Residential Sector				
Sub-total	28.73	33.56	41.60	51.25
Coal	11.61	11.28	10.74	10.09
Oil	4.27	5.14	6.61	8.37
Gas	3.64	4.83	6.81	9.18
Electricity	9.22	12.30	17.44	23.61
Total				
Sub-total	227.63	248.50	264.13	272.01
Coal	147.59	146.48	135.94	122.84
Oil	21.27	25.32	29.45	33.29
Gas	8.95	13.67	19.51	24.69
Electricity	48.84	61.81	77.85	89.60

Data source: calculated by the authors.

Table A2. End-use energy consumption demand up to 2030 in coordinated scenario (million tce).

	2015	2020	2025	2030
Primary Industry				
Sub-total	5.59	6.26	6.75	6.99
Coal	1.31	1.80	2.35	3.01
Oil	2.87	3.03	3.04	2.81
Gas	0.00	0.00	0.00	0.00
Electricity	1.20	1.18	1.07	0.81
Secondary Industry				
Sub-total	172.25	169.50	150.37	135.06
Coal	129.46	114.40	87.90	50.75
Oil	3.52	4.34	5.14	5.36
Gas	4.33	6.63	2.79	19.24
Electricity	34.95	44.13	54.54	59.70
Tertiary Industry				
Sub-total	21.06	28.05	34.23	41.42
Coal	5.21	6.94	8.47	10.25
Oil	10.61	14.13	17.24	20.87
Gas	0.98	1.30	1.59	1.92
Electricity	3.47	4.63	5.64	6.83

Table A2. *Cont.*

	2015	2020	2025	2030
Residential Sector				
Sub-total	28.73	33.56	41.60	51.25
Coal	11.61	11.28	10.74	10.09
Oil	4.27	5.14	6.61	8.37
Gas	3.64	4.83	6.81	9.18
Electricity	9.22	12.30	17.44	23.61
Total				
Sub-total	227.63	237.37	232.95	234.72
Coal	147.59	134.42	109.46	74.10
Oil	21.27	26.64	32.03	37.41
Gas	8.95	12.76	11.19	30.34
Electricity	48.84	62.24	78.69	90.95

Data source: calculated by the authors.

Table A3. Primary energy consumption demand up to 2030 in Hebei in two scenarios.

Scenario	Unit	Conventional Scenario				Coordinated Scenario			
		2015	2020	2025	2030	2015	2020	2025	2030
Primary energy consumption	million tce	336	391	439	471	336	380	407	431
By type									
Coal	million tce	260	249	249	249	260	207	166	125
Oil	million tce	28	40	47	50	28	40	47	50
Gas	million tce	11	40	47	53	11	40	67	86
Renewable energy	million tce	10	18	21	19	10	14	14	16
Nuclear energy	million tce	0	9	28	40	0	34	48	67
Transferred electricity	million tce	27	36	48	60	27	45	66	87
Coal	%	77	64	57	53	77	55	41	29
Oil	%	8	10	11	11	8	10	12	12
Gas	%	3	10	11	11	3	11	16	20
Renewable energy	%	3	5	5	5	3	5	5	5
Nuclear energy	%	0	2	6	9	0	9	14	15
Transferred electricity	%	8	9	11	13	8	12	16	20

Data source: calculated by the authors.

Table A4. Primary energy consumption and accounted CO_2 emission up to 2030 in Hebei in two scenarios.

Item	Unit	Conventional Scenario				Coordinated Scenario			
		2015	2020	2025	2030	2015	2020	2025	2030
Primary energy consumption	million tce	336	391	439	471	336	380	407	431
Accounted CO_2 emission	million ton	723	766	788	800	723	659	604	528
CO_2 emission intensity of primary energy consumption	ton/tce	2.15	1.96	1.79	1.70	2.15	1.74	1.48	1.22
The share of coal in total primary energy consumption	%	77	64	57	53	77	55	41	29
The share of non-fossil energy in total primary energy consumption	%	3	7	11	13	3	13	15	19

Data source: calculated by the authors.

References

1. National Bureau of Statistics, China. *China Statistical Yearbook*; China Statistics Press: Beijing, China, 2015.
2. Chen, H.X.; Li, G.P. Empirical Study on Effect of Industrial Structure Change on Regional Economic Growth of Beijing-Tianjin-Hebei Metropolitan Region. *Chin. Geogr. Sci.* **2011**, *6*, 708–714. (In Chinese) [CrossRef]
3. Lian, J.T. Research on the Economic Strategy of Hebei Province in the Coordinated Development of Beijing, Tianjin and Hebei. Ph.D. Thesis, Dongbei University of Finance and Economics, Dalian, Liaoning, China, December 2015. (In Chinese)
4. Liu, H.Y.; Xiao, L.; Xu, X.L.; Ban, Z.J. Analysis of Energy Consumption and Influence Factors of Carbon Emissions in Hebei Province. *J. Agric. Sci.* **2016**, *20*, 81–84. (In Chinese)
5. Zhang, J.W.; Gong, X.G. Research on the impact of industry structure change on energy consumption based on a case study of Hebei Province. *J. Yanshan Univ.* **2010**, *11*, 106–110.
6. Bureau of Statistics of Hebei, Hebei, China. *Hebei Economic Statistical Yearbook*; China Statistics Press: Beijing, China, 2016.
7. Du, Y.; Xiao, R.G.; Zhao, H.Q.; Fan, J.; Zhu, C.H. Dynamic analysis of the relationship between energy consumption and economic growth in Hebei: Based on the state space model. *China Min. Mag.* **2015**, *21*, 42–47. (In Chinese)
8. Department of Energy Statistics; National Bureau of Statistics, China. *China Energy Statistical Yearbook 2016*; China Statistics Press: Beijing, China, 2016.
9. Bureau of Statistics of Hebei, Hebei, China. Statistical Communique of Hebei on the 2014 National Economic and Social Development. Available online: http://info.hebei.gov.cn/hbszfxxgk/329975/330000/330496/6398050/index.html (accessed on 15 June 2017).
10. Institute of Tsinghua University at Hebei (ITUH). *Coordinated Development of Beijing, Tianjin and Hebei: Economic Development, Energy Consumption and Pollution Emission in Hebei Province*; ITUH: Beijing, China, 2015. (In Chinese)
11. Wu, Y.Q.; Zhao, Y.N. Beijing-Tianjin-Hebei Energy Consumption, Carbon Emissions and Economic Growth. *Econ. Manag.* **2014**, *28*, 5–12.
12. Bo, W.G.; Chen, F. The Coordinated Development among Beijing, Tianjin and Hebei: Challenges and Predicaments. *J. Nankai Univ.* **2015**, *1*, 110–118.
13. Zhang, G.; Wang, S.Q.; Liu, S.; Jia, S.J. Study on Co-ordination between Beijing-Tianjin-Hebei Based on Industry Matching and Transfer. *Econ. Manag.* **2014**, *28*, 14–20.
14. Wang, Z.H.; Lin, Y. Delinking indicators on regional industry development and carbon emissions: Beijing–Tianjin–Hebei economic band case. *Ecol. Indic.* **2015**, *48*, 41–48. [CrossRef]
15. Xinhuanet. Special Focus on Beijing-Tianjin-Hebei Cooperation and Development Plan. Available online: http://www.xinhuanet.com/syzt/jjjxtfz/index.htm (accessed on 27 June 2017).
16. Xinhuanet. The Central Government of China Decides to Set up Xiong'an New District. Available online: http://news.xinhuanet.com/politics/2017-04/01/c_1120741571.htm (accessed on 27 June 2017).
17. Zhu, E.J.; Wen, K.; Ye, T.L.; Zhang, G.X. *Annual Report on Beijing-Tianjin-Hebei. Metropolitan Region Development*; Social Science Academic Press: Beijing, China, 2016. (In Chinese)
18. International Energy Agency and Nordic Energy Research. Nordic Energy Technology Perspectives 2016 (Paris: OECD, 2016). Available online: www.iea.org/etp/Nordic (accessed on 15 June 2017).
19. Liu, Q.; Chen, Y.; Tian, C.; Zheng, X.-Q.; Li, J.-F. Strategic deliberation on development of low-carbon energy system in China. *Adv. Clim. Chang. Res.* **2016**, *7*, 26–34. [CrossRef]
20. Yuan, J.; Xu, Y.; Hu, Z.; Zhao, C.; Xiong, M.; Guo, J. Peak energy consumption and CO_2 emissions in China. *Energy Policy* **2014**, *68*, 508–523. [CrossRef]
21. Emodi, N.V.; Emodi, C.C.; Murthy, G.P.; Emodi, A.S.A. Energy policy for low carbon development in Nigeria: A LEAP model application. *Renew. Sustain. Energy Rev.* **2017**, *68*, 247–261. [CrossRef]
22. Lind, A.; Kari, E. The use of energy system models for analysing the transition to low-carbon cities—The case of Oslo. *Energy Strategy Rev.* **2017**, *15*, 44–56. [CrossRef]
23. Zhao, G.L.; Guerrero, J.M.; Jiang, K.J.; Chen, S. Energy modelling towards low carbon development of Beijing in 2030. *Energy* **2017**, *121*, 107–113. [CrossRef]
24. Pan, X.Z.; Chen, W.Y.; Clarke, L.E.; Wang, L.; Liu, G. China's energy system transformation towards the 2 °C goal: Implications of different effort-sharing principles. *Energy Policy* **2017**, *103*, 116–126.

25. Qiu, D.X.; Institute of Energy, Environment and Economy, Tsinghua University, Beijing. *Energy Planning and System Model*; Tsinghua University Press: Beijing, China, 1987. (In Chinese)

26. Niu, X.G.; Niu, J.G.; Dan, M. Energy Consumption and Carbon Emissions: Analysis and Prediction—The Case of Hebei Province in China. *Energy Procedia* **2011**, *5*, 2271–2277.

27. The Thirteenth Five-Year Plan of Hebei. Available online: http://www.hbdrc.gov.cn/web/web/t.xhtml?template=erji/fazhanguihua (accessed on 15 June 2017).

28. National Bureau of Statistics, China. Statistical Communique of the People's Republic of China on the 2014 National Economic and Social Development. Available online: http://www.stats.gov.cn/tjsj/zxfb/201502/t20150226_685799.html (accessed on 15 June 2017).

29. The Thirteenth Five-Year Plan of Hebei on Energy Conservation and Emission Reduction. Available online: http://info.hebei.gov.cn/eportal/ui?pageId=1962757&articleKey=6737879&columnId=329982 (accessed on 15 June 2017).

30. State Grid Energy Research Institute. *Analysis Report on Power Supply and Demand of China 2016*; China Electric Power Press: Beijing, China, 2016.

31. Dai, Y.D.; Bai, Q. *Scenario Analysis of Industrial Energy Conservation of China*; China Economic Publishing House: Beijing, China, 2015.

32. Wang, H.L. Model Simulation of China's Low-carbon Transportation Transformation Mechanism and Policy. Ph.D. Thesis, Tsinghua University, Beijing, China, 31 May 2016.

33. Institute of Energy Conservation in Buildings, Tsinghua. *2016 Annual Report on China Building Energy Efficiency*; China Architecture & Building Press: Beijing, China, 2016.

34. The Thirteenth Five-Year Plan of Hebei on Renewable Energy. Available online: http://info.hebei.gov.cn/eportal/ui?pageId=1966210&articleKey=6675503&columnId=330035 (accessed on 15 June 2017).

35. The Thirteenth Five-Year Plan of Hebei to Address Climate Change. Available online: http://info.hebei.gov.cn/eportal/ui?pageId=1966210&articleKey=6710248&columnId=330035 (accessed on 15 June 2017).

36. Nuclear Project Construction in Hebei Speed up—Nuclear Plants in Cangzhou Plans to be Constructed in 2016. Available online: http://www.china5e.com/news/news-889570-1.html (accessed on 15 June 2017).

37. Xiong, W.M. Development of the China Renewable Electricity Planning and Operations Model and Its Application. Ph.D. Thesis, Tsinghua University, Beijing, 6 September 2016.

38. Zhang, X. Development and Application of Regional Energy Emission Air-Quality Climate Health Model (REACH). Ph.D. Thesis, Tsinghua University, Beijing, China, 2016.

39. National Development and Reform Commission, CHINA. Revolutionary Innovation and Action Plan of the Energy Technologies. Available online: http://www.gov.cn/xinwen/2016-06/01/content_5078628.htm (accessed on 15 June 2017).

40. The State Council, CHINA. Strategic Action Plan of Energy Development. Available online: http://www.mlr.gov.cn/xwdt/jrxw/201411/t20141119_1335668.htm (accessed on 15 June 2017).

sustainability

MDPI

Article

Designing Sustainable Urban Social Housing in the United Arab Emirates †

Khaled Galal Ahmed

Architectural Engineering Department, College of Engineering, UAE University, Al Ain, P.O. Box 15551, UAE; kgahmed@uaeu.ac.ae; Tel.: +971-050-233-781

† This paper is an extended version of a paper titled "Thinking Beyond Zero-Energy Buildings: Investigating Sustainability Aspects of Two Residential Urban Forms in UAE" presented in the 5th International Conference on Zero Energy Mass Customized Housing-ZEMCH 2016, 20–23 December 2016, Kuala Lumpur, Malaysia.

Received: 30 June 2017; Accepted: 4 August 2017; Published: 10 August 2017

Abstract: The United Arab Emirates is experiencing a challenging turn towards sustainable social housing. Conventional neighborhood planning and design principles are being replaced by those leading to more sustainable urban forms. To trace this challenging move, the research has investigated the degree of consideration of sustainable urban design principles in two social housing neighborhoods in Al Ain City in Abu Dhabi Emirate, UAE. The first represents a conventional urban form based on the neighborhood theory; the other represents the new sustainable design. The ultimate aim is to define the obstacles hindering the full achievement of a sustainable urban form in this housing type. To undertake research investigations, a matrix of the design principles of sustainable urban forms has been initiated in order to facilitate the assessment of the urban forms of the two selected urban communities. Some qualitatively measurable design elements have been defined for each of these principles. The results of the analysis of the shift from 'conventional' to 'sustainable' case studies have revealed some aspects that would prevent the attainment of fully sustainable urban forms in newly designed social housing neighborhoods. Finally, the research concludes by recommending some fundamental actions to help meet these challenges in future design.

Keywords: sustainability; urban form; neighborhood; design; social housing; urban communities; UAE

1. Introduction

Sustainable urban form is generally defined as attaining sustainable urbanism through configuring its shape, function, and adaptability to change over time [1,2]. Handy and Niemeier [3] claim that sustainable urban form is relevant to residents' behavior within the built environment, thus stimulating residents to be more vigorous and to positively utilize urban spaces. Neuman [4] argues that sustainable urban form manifests both the process and the product that emerges from it. The formation of sustainable urban form integrates various sustainability features of cities in living, consuming, and producing. Accordingly, comprehending the influences of a specific urban form on environmental and social issues necessitates understanding the interrelated, inclusive, and adaptive processes producing such an urban form.

Additionally, one cannot define a single urban form as sustainable. Rather, there are various urban forms that suit a specific context—depending chiefly on the socioeconomic and environmental context of an area and its associated development objectives and plans. In general terms, an urban form can be called sustainable when it is responsive to the carrying capacity of the natural and the built environment, can provide a friendly living setting, and contributes to social justice [5].

Allen [6] outlines four dimensions that shape the sustainable urban form. First is the environmental dimension, which relates to the influence of urban production and consumption on the integrity and vigor of the urban area and its carrying capacity. Second is the social dimension, which entails the equity, inclusiveness, and adequacy of urban development, as this would endorse social justice that supports the livelihoods of residents in local communities. Third is the economic dimension, which entails the ability to exploit both local and regional resources for the welfare of the whole community. Fourth is the political dimension, which deals with the quality of urban governance in controlling the decision-making processes of different actors among the three former dimensions.

Social or public housing in the United Arab Emirates (UAE) has been developing since the establishment of the UAE in 1971. It started with the aim of providing public houses for nomadic and urban Emirati citizens. Currently, the main federal agency for the provision of social housing in the UAE is the Ministry of Public Works and Housing, established in 1972, and the Sheikh Zayed Housing Program, established in 1999 [7]. More recently, each Emirate has established its own local social housing agency to respond to its citizens' increasing demand for housing. These local social housing agencies include: Mohamed Bin Rashid Housing Establishment in Dubai, established in 2007; Sheikh Saud Housing Program in Ras Al Khaimah, established in 2008; the Housing Department in Sharjah, established in 2010; and Abu Dhabi Housing Authority, established in 2012 [8]. Since their establishment, both federal and local housing agencies have embarked on developing heavily subsidized social housing programs with a concentration on building social housing neighborhoods, following in their planning the principles of the conventional, functional, and self-contained Clarence Perry neighborhood design model. The areas of the housing plots in these neighborhoods have been generous, while the housing models were identical single-family houses that contained between three and five bedrooms on one or two floors.

In the last few years, the UAE has witnessed growing interest in sustainable development as the country has adopted a political agenda calling for achieving sustainability in all its development plans, including social housing. In accordance with that trend, the UAE has launched ambitious sustainable urban initiatives, such as ESTIDAMA (the Abu Dhabi Emirate's officially adopted version of the LEED (Leadership in Energy and Environmental Design), the internationally-recognized green building certification system developed by the U.S. Green Building Council (USGBC) and Masdar City, the first zero carbon emissions city in the world, to widen the application of sustainability in the building and urban housing sectors. In appreciation of this role, the UAE has been selected as the host country for the International Renewable Energy Agency (IRENA) headquarters [9].

This governmental policy of adopting a sustainable future agenda is reflected currently in social housing neighborhood design. Some limited pioneering projects have emerged lately in which conventional neighborhood planning and design principles are being replaced by what are perceived to be more sustainable ones.

2. Research Problem, Method, and Limitations

This research is concerned with investigating the 'level of consideration' of sustainable urban form design principles in the UAE's new social housing neighborhoods, which have been designated as 'sustainable' in comparison with conventional neighborhood designs. Al Ain City in Abu Dhabi Emirate has been selected as a locus for this investigation because it has been officially proclaimed as a model for green/sustainable cities in the UAE. To explore the research problem, a conceptual design matrix for the principles of sustainable urban form and their detailed 'design elements' was initiated through reviewing the relevant literature, as shown below, in order to facilitate the assessment of the produced urban forms of the two selected social housing communities in Al Ain: Al Salamat and Shaubat Al Wuttah, representing the past conventional and the present sustainability-orientated designs, respectively. Design principles of sustainable urban form and their relevant design elements have been defined in this conceptual design matrix.

In order to reach a satisfactorily qualitative assessment of these designs in the two case studies, two steps were conducted. First is qualitatively and individually assessing each of the defined design elements (as summarized in Table 1). Second is using the results of the assessments of each set of design elements as an indicator of the overall design principle associated with this set of design elements.

A simplified qualitative assessment scale was used in this assessment process, indicating five 'levels of consideration' for each of the design elements and, thus, their associated principles in the case study's urban form design, namely, 'fully considered', 'significantly considered', 'averagely considered', 'marginally considered', and 'not considered'. This assessment, of course, depends on the subjective interpretation of the analysis by the researcher and thus might not be fully accurate and might even be debatable, but the nature of the available data would make this qualitative assessment method the most appropriate. This limitation would mean that this research is actually measuring 'tendency' towards sustainable urban form in neighborhood design rather than accurately measuring each design element in detail, as this obviously would require further prolonged research, which is not the aim of this research. The investigation tools utilized in this qualitative assessment were CAD drawings, Google Earth maps, photographs, field observations during site visits, and informal interviews with Al Ain Municipality officials.

Comparing the assessment results of the new, sustainable neighborhood with those of the conventional one has helped with defining the changes and persistent challenges experienced in the design of the urban form of new neighborhoods. Ultimately, a proposed conceptual futuristic design scenario for achieving a more sustainable urban form of social housing in the UAE has been recommended through the interpretation of the outcomes of the research investigations. Figure 1 summarizes the research method, its utilized tools, and its intended outcomes.

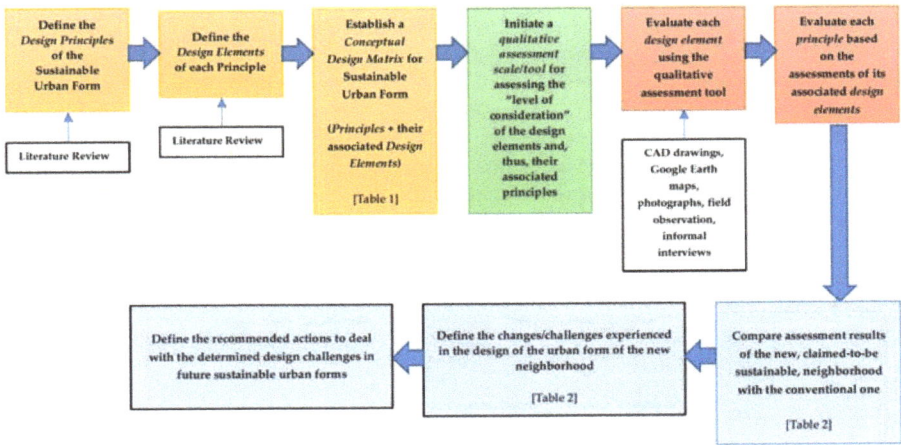

Figure 1. The Applied Qualitative Research Method and Its Tools.

3. A Conceptual Design Matrix for Sustainable Urban Form

To initiate a sustainable urban form matrix of principles and their relevant design elements, the research has consulted a wide variety of references about sustainable urbanism [5–7,10–15], New Urbanism [16], Smart Growth [17], Traditional Neighborhood Design (TND) [18], Transit-oriented Development (TOD) [19], Sustainable Neighborhood Planning [20], and Livable Communities [21]. With the acknowledged difficulty of developing a comprehensive set of design principles for sustainable urban form, given the various, mostly qualitative and intuitive definitions of such principles and their design elements, the main aim here was to collect the commonly agreed upon principles and design elements derived from the abovementioned resources. Still, this matrix might have unintentionally missed out some of the principles and/or relevant design elements. As detailed below, 12 design principles have been identified: Density/Compactness, Accessibility, Choice, Mobility,

Mixed use, Social mix/Social capital, Adaptability/Resilience, Local autonomy, Environmental quality, Community Safety and Security, Privacy, and Imageability/Sense of Place/Identity. These principles are highly interdependent. For example, appropriate density provides the population and activity basis for a sustainable neighborhood; mixed use and social mix shape the land use and social life in the neighborhood [20]. Also, walkable communities make pedestrian activity possible, thus expanding choice for transportation options, and creating a streetscape for a range of users including pedestrians, bicyclists, transit riders, and car drivers. To foster walkability, communities must mix land uses and build compactly, as well as ensure safe and inviting pedestrian corridors. Mixed land use also provides a more diverse and sizable population and commercial base for supporting viable public transit [17].

Finally, to avoid redundancy, the design element(s) related to a specific principle already mentioned in the matrix was not repeated, even if it also belongs to another one.

3.1. Principle 1: Density/Compactness

Several studies have indicated that urban sprawl encourages car-oriented lifestyles and consequently entails higher urban management costs, accompanied by intensive travel and associated negative environmental effects [13]. On a per-unit basis, it is more economically and environmentally sustainable to provide and maintain services like water, sewers, electricity, phone service, and other utilities in more compact neighborhoods than in dispersed communities [17]. Studies show that a doubling of density results in a 30% reduction in energy use per capita. The lower-density cities of the United States (typically 10 persons per hectare or less) use about five times more energy per capita in gasoline than the cities of Europe, which are in turn about five times denser on average. A compact city with good public transport, walkability, and a reduced need to drive long distances to reach destinations adds to environmental sustainability [21].

Design Elements: Jenks and Burgess [10] maintain that achieving a sustainable urban form entails developing a compact built-up area with appropriate population densities in order to boost various human activities. To promote a high population density, Frey [5] defines a minimum gross population density of 50 to 60 persons per hectare (pph) to be sufficient for supporting viable local services, facilities, and public transport. Compactness and concentration of urban functions within the urban area lead to environmental, social, and economic sustainability benefits. Urban intensification is pointed out as a major approach for achieving compactness and usually achieved through using urban land more efficiently by increasing the density of activities [11]. So, in such a compact urban form the neighborhood services core may have an area of about 1 ha [5]. Local provision of daily amenities, services, and facilities, including public open and green spaces, should be accessed by foot, bicycle, or a short public transportation ride [15].

3.2. Principle 2: Accessibility

Sustainable urban form is conventionally measured by how the urban form affects vitality, the degree to which the settlement form fits the requirements of its residents, and how able people are to access activities and services [12].

Design Elements: Neighborhood shared amenities and public transportation nodes should be located within walkable distances of houses [5,18]. Walkability in a neighborhood can be measured by the walking distance to key services, which is usually about 400 m [20] and increased to be about 600 m for the public transportation nodes [22]. It is also preferable to gather the local services and facilities centrally around the transport node because this will create a vivid and mixed-use central place. Furthermore, this would make the transport station work as a catalyst for great place-making around it [19]. Ease of access to local services and facilities should be guaranteed to less mobile citizens and those who do not drive, such as the elderly, children, and the disabled [16]. For public transport within the neighborhood, bus services should link neighborhood clusters and services with 200 m to 300 m stop intervals [20]. Additionally, neighborhood residents should have access to district and city centers through efficient public transport [5]. Public transport should have priority and exclusive lanes such as bus and tram lanes. Finally, the provision of a proper signage system will help guide users to their destinations easily [23].

3.3. Principle 3: Choice

Providing people with more choices in housing, shopping, communities, and transportation is a key aim of sustainable urbanism [17].

Design Elements: A hierarchy of amenities of different capacity and scale should be available for people to choose from [18]. Sustainable communities are seeking a wider range of transportation options in an effort to improve beleaguered current systems [17]. These options would allow people to choose their mode of mobility (cycling, walking, bus, or car). The catchment areas should overlap with other neighborhoods [5]. In addition, the provision of various designs of housing units and buildings can support a more diverse population and allow more equitable distribution of households of all income levels [17]. Actually, it is claimed that to qualify as sustainable housing, a project should include a range of housing types [18].

3.4. Principle 4: Mobility

Different modes of mobility should be provided in the sustainable neighborhood, including walkability, cycling, public transport, and private cars. The design should support a high degree of mobility through efficient types of transportation for accessing services through short trips. Encouraging transit use helps reduce air pollution and congestion [17].

Design Elements: Interconnected and hierarchical networks of streets should be designed to encourage walking, reduce the number and length of automobile trips, and conserve energy [16,20]. For boosting walkability, there is a need for a safe, pleasant, and lively environment that encourages walking to a station, shopping, and other services and facilities [19]. In such a hot climate, shaded pedestrian walkways are essential, as recommended in the Al Ain 2030 Plan [24]. For cycling, a safe network with well-distributed cycle parking spots leading to transport nodes, services, and facilities should be considered in neighborhood design [19]. For neighborhood public transport, Frey [5] emphasizes that the provision of public transport has been proven to be the most economical way to facilitate mobility in a city, which in turn necessitates a modular city context composed of urban 'cells' or 'proximity units'. The interrelationship of people, transport, and services is thus considered the core of the microstructure of the city. Therefore, the availability of efficient, coordinated, fast, comfortable, and inexpensive public transport (bus or LRT) providing access to district and city centers is essential for the design of a sustainable urban form [15]. For safety reasons, calming traffic inside neighborhoods should be considered through using road bumps and/or other traffic-calming measures [23]. On the other hand, car parking is to be exclusively allowed to locals, while no car parking should be allowed in neighborhood centers, except for vehicles for disabled people and taxis [5].

3.5. Principle 5: Mixed Use

The core concept of mixed land use is a critical component of achieving sustainable urban form. By putting a diversity of activities and services, such as residential, commercial, recreational, and other uses in close proximity to one another, alternatives to driving, such as walking or biking, become viable [17,20]. Jabareen [11] claims that there is a noticeable consensus among researchers that mixed use plays a vital role in realizing sustainable urban form. Heterogeneous zoning permits land uses to be located in close proximity to one another and thereby decreases the travel distances between activities.

Design Elements: Community planning provides integrated residential, commercial, recreational, and civic uses that are essential to the daily life of residents of differing demographic profiles and are connected by both public and private transportation options [18]. According to UN Habitat [20], the suggested floor area distribution for a sustainable neighborhood is: 40–60% for economic use, 30–50% for residential use, and 10% for public services. The set of recommended standards is a range to allow for flexibility so that different cities can adapt them to their own situations. Barton [15] suggested the multi-use of buildings for both commercial purposes and housing. For example, housing can be built over shops and service outlets. This will create places with multiple destinations within close proximity, where the streets and sidewalks balance multiple forms of transportation [17]. For mixed use to be effective, appropriate workplaces that do not cause harm to residents or the environment should be allowed [15].

3.6. Principle 6: Social Mix/Social Capital

Social mix aims to promote the cohesion of and interaction between different social classes in the same community and ensure accessibility to equitable urban opportunities. Social mix provides the basis for healthy social networks and social capital, which in turn are the driving force of city life. Social mix and mixed land-use are interdependent and promote each other. Mixed land-use and appropriate policy guidance lead to social mixing. In a mixed land-use neighborhood, job opportunities are generated for residents from different backgrounds and of different income levels. People live and work in the same neighborhood and form a diverse social network [20].

Design Elements: Supporting social mix requires the provision of housing plots in different sizes and with different regulations, to increase the diversity of housing options [20]. Within neighborhoods, a broad range of housing types and price levels can bring people of diverse ages, races, and incomes into daily interaction, strengthening the personal and civic bonds essential to an authentic community [16]. Providing quality housing for people of all income levels is an integral component of any sustainable urbanism strategy. By creating a wider range of housing choices, communities can mitigate the environmental costs of auto-dependent development, use their infrastructure resources more efficiently, ensure a better job–housing balance, and generate a strong foundation of support for neighborhood transit stops, commercial centers, and other services [17]. Therefore, provision of a wide range of dwelling types including single-family houses and low-rise apartment buildings in addition to a wide range of tenure types (government, ownership, and rent) are essential pillars for realizing a sustainable urban form [20]. On the other hand, a range of parks, from tot lots and village greens to ballfields and community gardens, should be distributed within neighborhoods [16]. These and other social activity nodes will trigger community interaction, create social networks, and increase social capital. Accordingly, community life will be revitalized by making open spaces, green spaces, and pedestrian-oriented retail and social nodes where people meet [17].

3.7. Principle 7: Adaptability/Resilience

The economic health and harmonious evolution of neighborhoods, districts, and corridors can be improved through graphic urban design codes that serve as predictable guides for change [16].

Design Elements: On both the housing unit/building and urban scales, sustainable urban design should provide the ability to change and expand in response to changing socioeconomic conditions without major upheaval [15].

3.8. Principle 8: Local Autonomy

The common thread, however, is that the needs of every community and programs to address them are best defined by the people who live and work there. Citizen participation can be time-consuming, frustrating, and expensive. Nonetheless, encouraging community and stakeholder collaboration can lead to creative, speedy resolution of development issues and greater community understanding of the importance of good planning and investment. Sustainable urban form plans and policies developed without strong citizen involvement will lack staying power. Involving the community early and often in the planning process vastly improves public support for sustainable urban form and often leads to innovative strategies that fit the unique needs of a particular community [17].

Design Elements: The neighborhood forms an identifiable area that encourages citizens to take responsibility for their maintenance and evolution [16]. Sustainable urban form is conventionally measured by how much control people have over services, activities, and urban spaces [12]. The ability of individuals and the community as a whole to shape their own environment and take part in the decision-making process within the local community is vital. Sustainable urban form encourages communities to craft a vision and set standards for development that respect the community values of architectural beauty and distinctiveness, as well as expand choices in housing and transportation [17]. Local production of food and utilization of renewable energy resources in the neighborhood are two other dimensions of local autonomy [15].

3.9. Principle 9: Environmental Quality

All the above principles and their design elements will enhance the environmental quality of the neighborhood through increasing energy efficiency and decreasing pollution.

Design Elements: In addition to the previous design elements, having private and semi-private green areas for residential units can enhance the environmental quality [5]. Encouraging the planting of public open spaces and streets is another important design element towards realizing this goal [17] as the preservation of open space benefits the environment by combating air pollution, attenuating noise, controlling wind, providing erosion control, and moderating temperatures [17].

3.10. Principle 10: Community Safety and Security

Besides the abovementioned design factors (safe pedestrian walkways and traffic-calming measures), sustainable urban design should cater to community security and consider safety measures for all.

Design Elements: Considering visual surveillance in the public realm through the urban design of the housing plots and other buildings is an essential design measure [23]. Mixed use and high density can enhance the vitality and perceived security of an area by increasing the number and activity of people on the street [17]. Safety for all members of the community can be achieved through inclusive design that considers the needs of children, the disabled, and elderly people [5].

3.11. Principle 11: Privacy

Privacy is an essential sociocultural aspect in sustainable urban design in Al Ain [24].

Design Elements: the provision of a personal private outdoor space for each house with gardens, roof gardens, and terraces helps to ensure privacy for the residents [23].

3.12. Principle 12: Imageability/Sense of Place/Identity

Dempsey et al. [14] mentioned that community stability and a sense of belonging to a place were found to be influenced not only by non-physical aspects such as feelings of satisfaction with the neighborhood, but also by a number of physical features including: density, type of accommodation and its location in relation to surrounding services and facilities, public transport, and the city center. Sustainable urban design creates interesting, unique communities that reflect the values and cultures of the people who reside there, and fosters physical environments that support a more cohesive community fabric. It promotes development that uses natural and manmade boundaries and landmarks to define neighborhoods. Communities are able to identify and utilize opportunities to make new developments conform to their standards of distinctiveness and beauty [17].

Design Elements: The design of spaces and structures should reflect and celebrate what is unique about a community's people, culture, heritage, and natural history [18]. This would be reflected in the urban and architectural design of buildings, streets, streetscapes, green spaces, plantings, etc. [15].

Finally, Table 1 summarizes the conceptual design matrix for sustainable urban design including the 12 principles and their quantitative/qualitative measurable design elements. As discussed later, this matrix was used for assessing the 'level of consideration' for the sustainable urban form principles in the two social housing schemes of Al Salamat and Shaubat Al Wuttah.

Table 1. A Conceptual Design Matrix for Sustainable Urban Form (Sources: References in Section 3).

Principal	Design Elements
Density/Compactness	• Gross population density of 50 to 60 pph. • Compact service core with an area of about 1 ha. • Local provision of daily amenities, services and facilities, including public open and green spaces, accessed by foot, bicycle, and short public transportation ride.
Accessibility	• Shared amenities and public transportation nodes should be located within maximum walkable distances of 400 m and 600 respectively from houses. • Ease of access to local services and facilities by less mobile category. • Local services and facilities centrally gathered around the transport node. • Buses linking neighborhood clusters and services together with 200 m to 300 m stop intervals. • Efficient public transport providing access to district and city centers. • Public transport has traffic priority and exclusive lanes. • Proper signage system.
Choice	• A hierarchy of services and facilities of different capacity and scale. • Choice of mode of mobility. • Catchment areas overlap with other neighborhoods. • Various housing units/buildings in sort and design.
Mobility	• Interconnected and hierarchical networks of streets. • Safe, shaded, well-lit and pleasant pedestrian routes leading to transport nodes, services and facilities. • Safe bikeway network with well-distributed cycle parking spots leading to transport nodes, services and facilities. • Availability of efficient, coordinated, fast, comfortable, and inexpensive public transport providing access to district and city centers. • Calmed street traffic inside neighborhood. • Car-parking for locals and no car parking at neighborhood centers except for vehicles of disabled people and taxis.
Mixed Use	• Provision of integrated residential, commercial, recreational, and civic uses that are essential to the daily life of residents of differing demographic profiles and are connected by both public and private transportation options. • 40 to 60 per cent of the floor area for economic use, 30 to 50 per cent for residential use and 10 per cent for public services. • Multi use of buildings for commercial and housing. • Allowing for appropriate workplaces.
Social Mix/Social Capital	• Plots in different sizes and with different regulations, to increase the diversity of housing options. • Wide range of dwelling types. • Wide range of tenure types. • Social activities nodes including recreational facilities.
Adaptability/Resilience	• Ability to change and expand in response to changing socio-economic conditions without major upheaval. • Adaptability should be achieved on both the housing unit/buildings and urban scales.
Local Autonomy	• Ability of individuals and the community to shape their own environment and take part in the decision making within the local community. • Production of food. • Utilizing renewable energy resources.
Environmental Quality	• Private and semi-private green areas for residential units. • Plantation of public open spaces and streets.
Community Safety and Security	• Visual surveillance in the public realm. • Inclusive design that considers the needs of children, the disabled and elderly people.
Privacy	• Private outdoor space for each house with gardens, roof gardens and terraces.
Imageability/Sense of Place/Identity	• Distinguishable features and sets of activities. • Distinguished identity of neighborhoods in urban and architectural design.

4. Results

4.1. Analysis of the Urban Form Design Principles of Al Salamat: The Conventional Past Design Experience

The aim of this analysis is to investigate the sustainability gaps in the conventional urban form of social housing neighborhoods that recent designs are envisaged to bridge. Al Salamat is a typical social housing neighborhood that was developed in 2000. It is located approximately 21 km west of Al Ain city center (Figure 2).

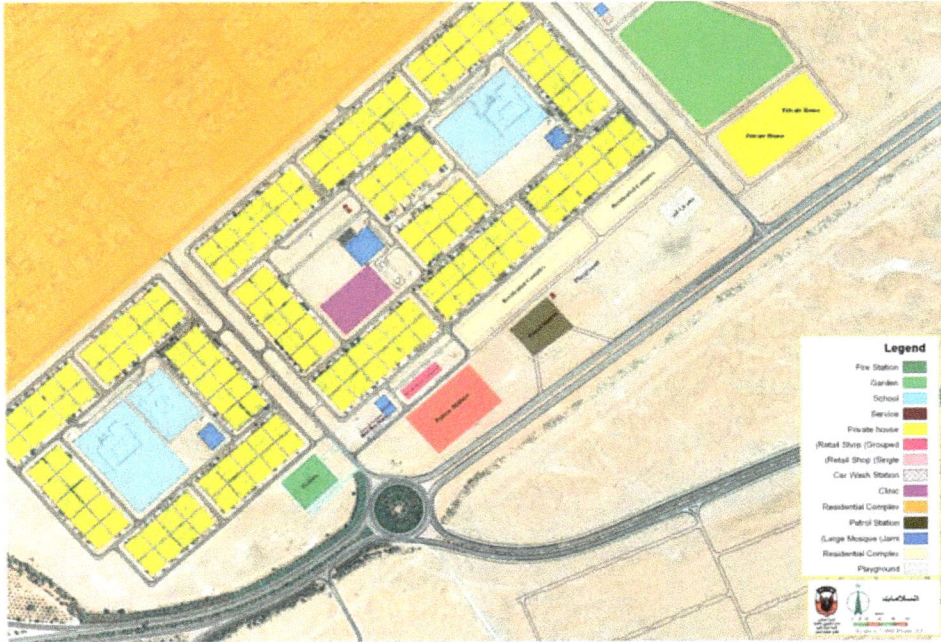

Figure 2. Al Salamat Community and Its Planned Land Use. (Source: Google Earth).

4.1.1. Density/Compactness

The site area is about 132.5 ha. It contains 166 single family houses with plot dimensions of 45 m × 45 m. The gross population density is about 11.3 pph, which is a remarkably low density that mainly resulted from the adopted sprawl design of a dominantly car-oriented urban form (Figures 2 and 3). The neighborhood has fragmented service in six zones: three of them are located in the middle of the three residential blocks, and the other three are located on the neighborhood edges, with a total area that far exceeds the suggested compact service core of 1 ha (Figure 2). Many of the amenities, services, and facilities that are needed daily are not locally provided and those that are provided, as discussed in the following sections, are not easily accessible for many houses except by private cars. Accordingly, the results of the assessment of these design elements make this principle "Marginally Considered".

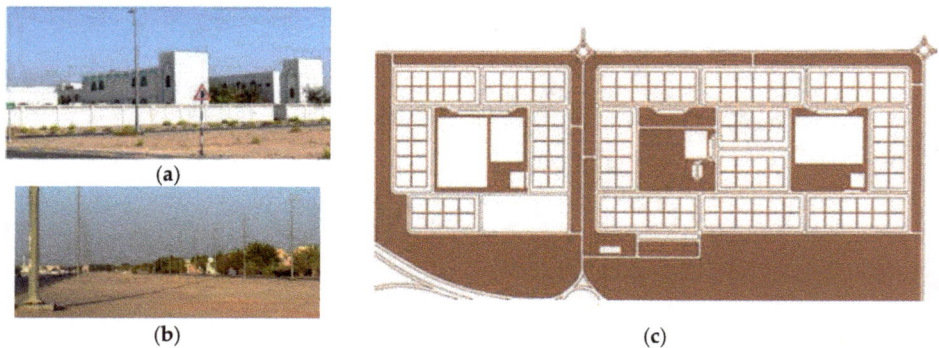

Figure 3. (a–c) Low density and Sprawl Design of Al Salamat.

4.1.2. Accessibility

In Al Salamat, as illustrated in Figure 2, only one nursery serves a residential block and the primary schools serve only one and half blocks. Students who live by the edge of the neighborhood need to walk almost double the standard distance (1.25 km). Private cars and school buses are commonly used to transport students. Middle schools are located in other neighborhoods about 1.8 km away. Meanwhile, secondary schools are found in Al Yahar District, about 7 km away. Retail shops are on the edge of the neighborhood and beside each mosque within the residential blocks, which allows for multi-purpose trips. Other planned services like clinics, playgrounds, and parks have not been developed yet, apparently because of the low population density, which negatively affects the economic vitality of the neighborhood. It is also noticed that some inappropriate locations of the internal crossroads cause longer trip lengths.

No accessibility measures were considered for the less mobile categories of residents. The location of the local services and facilities has nothing to do with the two bus stops (transportation nodes). The bus route does not link the neighborhood clusters and the two bus stops are about 600 m apart, which is almost double the standard distance. The bus line connects the neighborhood to the city center. There are no priority or exclusive lanes for buses. The signage system is proper. The results of the assessment of these design elements point out that this principle is "Marginally Considered".

4.1.3. Choice

It is noticed that the existing services and facilities in Al Salamat are limited in number and quality (Figure 4). There is a lack of pedestrian walkways and bus stations, while cycling lanes are totally absent. The neighborhood was originally designed to have more services such as a clinic, public park, and more retail shops that, if developed, would help improve choice.

The residents in Al Salamat depend on services in adjacent neighborhoods such as schools, mainly using private cars to reach these services. Also, there is no variety of housing types in this neighborhood. Only some inhabitants managed to change/extend their houses over time. The originally proposed master plan has assigned zones for residential apartment blocks that would, if implemented, enrich housing diversity and create more choices. The results of the assessment of these design elements make this principle obviously "Marginally Considered".

Figure 4. Limited Choices of Services and Utilities in Al Salamat. (Source: Google Earth & the Author).

4.1.4. Mobility

Al Salamat has interconnected and hierarchical networks of streets prepared only for private car traffic. There are some connections with surrounding neighborhoods, but mainly via private cars. It has been noticed that many residents use informal routes, driving in sandy open spaces to minimize their trip lengths. The inappropriate catchment distances compel residents to use their own cars and the unsuitable locations of some services require longer trip lengths. Additionally, the streets of Al Salamat are generally risky, with narrow and frequently interrupted sidewalks. The landscape of the neighborhood is quite poor, so there are no sitting or walking facilities. Also, there are no cycling paths and there are four generous parking lots for each housing unit (Figure 5).

For public transportation, sustainable neighborhood design should cater to public transportation nodes and provide fewer parking spaces. As shown in Figure 5, only two bus lines are serving the whole neighborhood and the locations of the bus stops are not easily accessible by most residents. Moreover, there is no appropriate shading for the bus stop. Bus timings are not shown in many cases. The availability of generous car parking lots everywhere makes the use of private cars more appealing. Finally, the results of the assessment of these design elements indicate that this principle is "Marginally Considered".

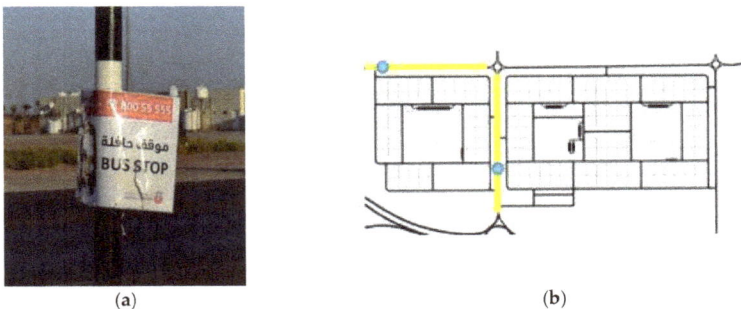

(a)

(b)

Figure 5. *Cont.*

Figure 5. Mobility Modes in Al Salamat, (**a**,**b**) Insufficient Public Transportation Nodes; (**c**) Lack of Pedestrian and Cycling Lanes and (**d**) Car-oriented Mobility.

4.1.5. Mixed Use

Mixed use encourages developing local offices/workshops, home-working, and multiple uses of space. It also helps diversify accessible job opportunities with good local training services. While the original master plan of Al Salamat (Figure 2) provides more mixed use opportunities, the actual as-built status has less mixed use as it lacks many services and working opportunities (Figure 4). Multi-use of buildings is not facilitated because the only housing type in this neighborhood is single-family housing. The results of the assessment of these design elements render this principle "Marginally Considered".

4.1.6. Social Mix/Social Capital

In Al Salamat, housing plots are identical in size and regulations, leaving no room for diversity. As mentioned above, all housing types are single-family housing units with only one typical model. A single tenure type is adopted where all houses are government-designed and delivered to citizens on a low income. No public space is provided for events. In the original master plan of the neighborhood, parks are located in the periphery while the actual active park is far from the neighborhood and only serves families. The lack of sitting and walking facilities decreased residents' casual meetings. Accordingly, the results of the assessment of these design elements make this principle "Marginally Considered".

4.1.7. Adaptability/Resilience

Adaptability is crucial for any sustainable urban design where buildings are designed for use change and houses are designed to be easily extended for evolving family circumstances while open spaces should permit a variety of social activities. Due to their ample area of 45 m × 45 m, housing plots allowed for the extension of private houses and many residents have embarked on changes that responded to their needs. On the other hand, open spaces, which are noticeably deserted with no clear definition or ongoing social activity, could be multi-functional and adapted to different socioeconomic activities if appropriately utilized. The results of the assessment of these design elements indicate that this principle is "Averagely Considered".

4.1.8. Local Autonomy

There was no participation in the design with local authorities. They are not involved in managing and/or maintaining local community resources such as parks, sport fields, etc. Production of food in the green areas and private green spaces is not noticed except by a few houses. Renewable energy resources are not utilized for locally producing required energy, even partially. The results of the assessment of these design elements make this principle "Not Considered".

4.1.9. Environmental Quality

Despite the presence of some private green areas within housing plots and semi-private green areas developed by the residents themselves in front of the housing plot, the environmental quality in Al Salamat is adversely affected by heavy reliance on private cars due to inappropriate connectivity and long walking distances to services and facilities, an inefficient public transportation system with a few bus stops with no proper shade, inappropriate pedestrian sidewalks, and the lack of utilization of renewable energy sources. Public open spaces are minimal. The results of the assessment of these design elements reveal that this principle is "Averagely Considered".

4.1.10. Community Safety

As mentioned above, in Al Salamat, sidewalks are narrow and unsafe and there are no cycling paths, while many linear streets are without humps or other traffic-calming measures. Many residents use shortcuts, driving in open spaces to minimize their trip lengths (Figure 6b). It is noticed that the spaces between plots are not utilized as pedestrian walkways seem risky to use (Figure 6a). Speed control traffic signs are well distributed within the neighborhood. Furthermore, the inappropriate catchment distances minimize social interaction and thus reduce opportunities for visual surveillance, which has been adversely affected by the solid high fences of the houses that are facing the public (Figure 6a). Principles of inclusive design that consider the needs of children, the disabled, and elderly people are significantly absent. On the other hand, there is noticeable social homogeneity among neighbors, with very low rate of crimes and social problems. The results of the assessment of these design elements render this principle "Averagely Considered".

(a)   (b)

Figure 6. Safety Issues in Al Salamat: (**a**) in-between Spaces among Fences of the Housing Plots; (**b**) Shortcuts Made by Private Car Drivers through Unutilized Open Spaces.

4.1.11. Privacy

The design provides amble private space within each housing plot, with wide gardens. Roof gardens and terraces are not favored by the residents (Figure 3). This clearly indicates that the privacy principle is "Significantly Considered" in the design of the neighborhood.

4.1.12. Imageability/Sense of Place/Identity

No distinguishable features or social activities in the open spaces have been noticed. Actually, little effort has been expended to identify distinctive architectural and urban characteristics for buildings and urban spaces in Al Salamat, as can be easily noticed. This assessment renders this principle "Marginally Considered".

The overall results of the analysis of the urban form design principles of the Al Salamat neighborhood are summarized in Table 2.

4.2. Analysis of the Urban Form Design Principles of Shaubat Al Wuttah: The Present Turn towards Sustainable Design

The main aim of the analysis of this recently designed and currently under development Shaubat Al Wattah social housing local community is to investigate to what extent its sustainability-orientated design has managed to bridge the sustainable urban form gaps found in Al Salamat, a conventional neighborhood. Shaubat Al Wattah is located on the southeast side of Al Ain City, about 15 km from the city center. It covers a total area of about 460 ha, encompassing 1580 single family houses in five housing clusters, as shown in Figure 7. Each housing cluster center contains a mosque, a kindergarten, some retail shops, and a playground.

4.2.1. Density/Compactness

The total targeted population is about 14,000 persons. The housing plot area is 30 m × 36 m, which is much smaller than in the conventional case study of Al Salamat (45 m × 45 m). This reflects a tendency towards a more compact and dense urban form. The gross population density is about 30.5 pph, which is remarkably higher than the gross population density of the conventional case study (11.3 pph). So, from a comparative point of view, the urban design of this neighborhood is more compact and much denser than the conventional one. However, it is still significantly less than the global standard of 50 to 60 pph. On the other hand, besides the five small housing clusters service areas (Figure 7), there is a planned mall located in the upper left corner in an area apparently not easily accessible by many residents, as discussed below.

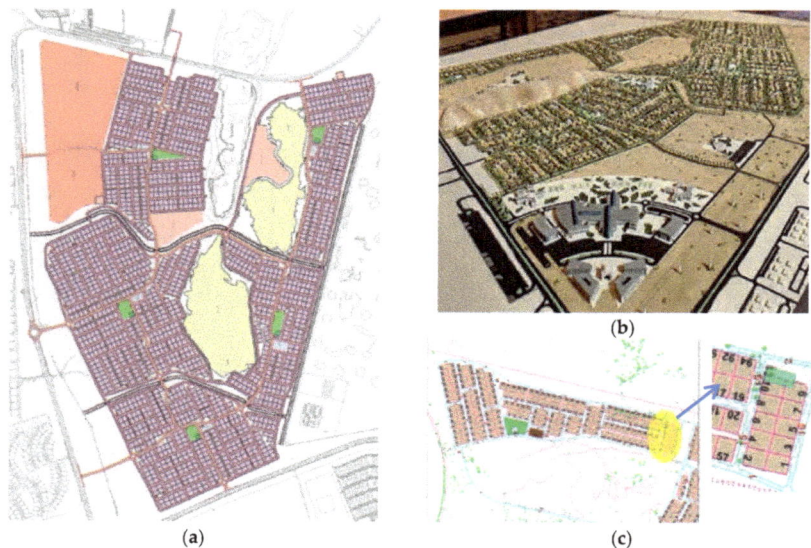

Figure 7. Shaubat Al Wuttah Community: (**a**) Plots and Services; (**b**) Perspective Showing the Designed Community Services; and (**c**) Arrangement of Housing Plots. (Source: Al Ain Municipality).

The results of the assessment of these design elements indicate that the principle of Density/Compactness is "Marginally Considered".

4.2.2. Accessibility

The decentralized distribution of the services and facilities needed daily made them accessible for many houses due to the adopted compact pattern of the urban form. Comparatively, the urban design of this case study has achieved better accessibility measures than the conventional one. Still, the neighborhood urban design does not fully respect the standard catchment areas as various services and facilities are not accessible by walking. As an example, the upper right side housing cluster (Figure 8) shows that the solely developed nursery is accessible by walking to about 12 h only, i.e., only 10% of the total villas in this cluster. Also, the playground area is accessible for about 70% of the houses in the same cluster. Meanwhile, the primary and secondary schools are located in far areas in the neighborhoods, thus reaching them necessitates long trips using private cars or buses. This applies to other clusters in one way or another.

Accessibility for less mobile categories has been included in the design such as wheelchair requirements at traffic intersections and so on. The urban design does not consider the high concentration of local services and facilities around the transport node (bus stops). Theoretically, buses routes are going to link the neighborhood clusters and services together with stop intervals of about 300 m. The proposed efficient bus routes are supposed to provide access to Al Ain city center. Still, public transport has no traffic priority or exclusive lanes. Additionally, a proper signage system is expected as is the case in the whole city of Al Ain. Accordingly, the results of the assessment of these design elements make this principle "Averagely Considered".

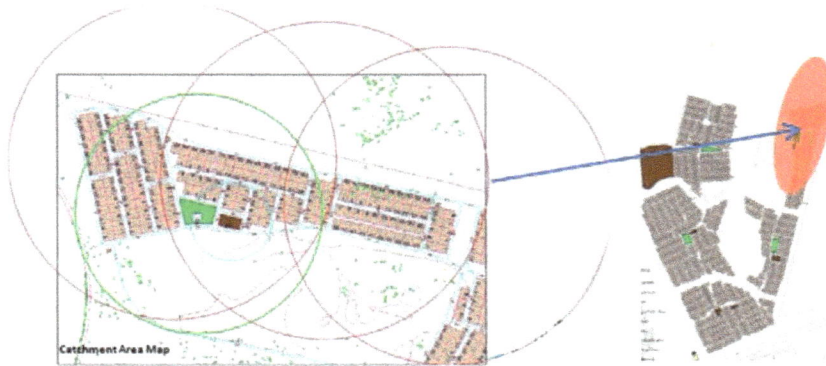

Figure 8. Catchment Areas in a Housing Cluster in Shaubat Al Wuttah Local Community. (Source: Al Ain Municipality).

4.2.3. Choice

There are limited options for services and facilities in this local community. Therefore, the reliance on private cars is high. Moreover, the neighborhood lacks many services including health centers, clinics, and private schools that would meet residents' diversified needs. Unlike Al Salamat, the design here provides attractive pedestrian walkways and cycling lanes within a safe environment and distributed rest areas among walking paths. This would enable a choice between modes of travel. The catchment areas do not overlap with other neighborhoods and residents, similar to those of Al Salamat, and people will tend to use their private cars to reach services and facilities missing in their proximity.

Diversification of housing types and sizes has been considered, but only for single-family houses, where nine design models were made available for residents to select from. The housing models reflect three different architectural styles, namely Islamic, Modern, and Traditional (Figure 9). Some house models have "stepless" entrances and other accessible features so that the handicapped and/or elderly people can select those designs. Other housing types, such as apartments, are not provided. So, the results of the assessment of these design elements point out that the Choice principle is "Averagely Considered".

Figure 9. Architectural Styles of the Housing Models in Shaubat Al Wuttah Local Community Project. (Source: Al Ain Municipality).

4.2.4. Mobility

The design of Shaubat Al Wuttah has achieved good integration between housing clusters within the neighborhood, housing clusters, and neighborhood services, services in each neighborhood, and between the neighborhood and the city (Figure 7). In accordance with the Al Ain 2030 Plan [24], the urban design of Shaubat Al Wuttah provides attractive pedestrian walkways within a safe environment and rest areas carefully distributed among walking paths. This is envisaged to encourage walking as an alternative mode of transport. Cycling paths are provided in the design as another alternative. Figure 10 shows a conceptual design for a primary collector road in Shaubat Al Wuttah with dedicated safe, shaded, well-lit, and pleasant pedestrian and cycling lanes. Nonetheless, as discussed above, the inappropriate catchment distances would make walking and cycling less favorable options and rather would encourage private car reliance as the primary mobility mode for residents.

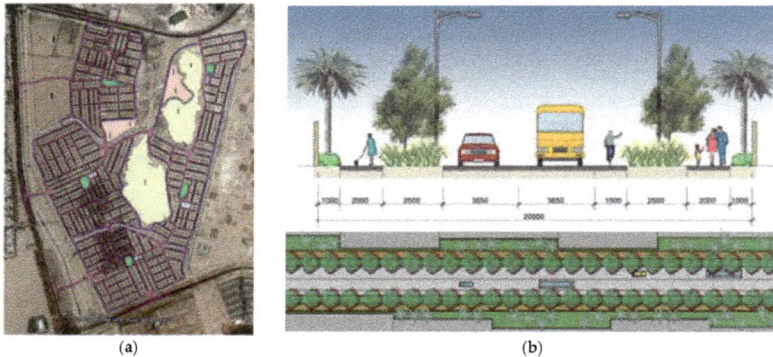

(a)   (b)

Figure 10. (**a**) Road Network; (**b**) Section and Plan for A Primary Collector Road in Shaubat Al Wuttah. (Source: Al Ain Municipality).

There are bus stops planned, but they are not clearly defined yet. The streets would have traffic-calming measures. This integrated system of different modes of travel within the neighborhood's road network, including pedestrian walkways, cycling paths, and bus stops, as shown in Figure 10, makes a big difference compared to the conventional neighborhood form. Furthermore, in the neighborhood design car parking lots have been decreased to promote the use of public transport. Actually, the number of parking lots exceeds 3160, which is still a considerable number and would not encourage residents to abandon using their own cars. The results of the assessment of these design elements make this principle "Significantly Considered".

4.2.5. Mixed Use

The neighborhood design does not provide all the required services and facilities of daily life. Many of the designed services and facilities are not within the standard catchment area, which inevitably increases reliance on private cars. The neighborhood design does not provide suitably diverse and accessible job opportunities on the neighborhood level such as offices/workshops and, hence, the design is not catering for multi-function spaces and multi-function trips. The recommended areas of different mixed uses have not been achieved as residential use is still dominant, even with the introduction of the mall by the edge of the site. Similar to the Al Salamat conventional neighborhood, the multi-use of buildings for both commercial and housing purposes by, for example, providing housing over shops and service outlets is not considered here. Appropriate work places are not allowed either. The results of the assessment of these design elements render the Mixed Use principle "Marginally Considered".

4.2.6. Social Mix/Social Capital

Housing plots here, as in Al Salamat, are identical in size, with no change in regulations. Meanwhile, as mentioned above, there are nine different housing unit models that residents can select from. The housing tenure type is similar to that of Al Salamat, where the government builds and delivers the houses to citizens with a low income. The provision of safe and attractive pedestrian walkways with landscaping and distributed rest areas in the design of Shubat Al Wuttah does not seem to actually encourage walkability and encounters between people, mainly due to the inappropriate catchment areas and the lack of diversified services and facilities, as mentioned earlier. Furthermore, the neighborhood design lacks sufficient public open spaces. There is only a playground area with a mosque within it, which minimizes opportunities for residents' participation in any shared collective activities that might take place in such open public spaces. This, of course, will not encourage social relations and interactions among residents.

Based on the results of the assessment of these design elements, this principle can be considered "Marginally Considered".

4.2.7. Adaptability/Resilience

For adaptability and resilience of the design models of single-family houses, originally there were only four different designs but they have been increased to nine in response to different user needs. Most designs have options for limited future extension. Figure 11 illustrates two examples of pre-designed expansions in the first floors of two house models.

(a)   (b)

Figure 11. Two Typical House Models in Shaubat Al Wuttah Urban Community: (**a**) Ground Floor and (**b**) First Floor Plans. The Expandable Area is Shown in Dark Gray on the First Floor. (Source: Al Ain Municipality).

Moreover, the unified plot area of Shaubat Al Wuttah was reduced to 30 m × 36 m, which made it impractical to extend it on the ground level. This obviously does not respond to diversified residents' needs as they are forced to comply with the fixed pre-designed extensions. For adaptability on the urban scale, the provided roads, paths, and narrow squares do not actually cater for adaptive social and economic activities. The only vacant space in the middle of the development is the foothills of Hafeet Mountain (Figure 10). Accordingly, the results of the assessment of these design elements make this principle "Marginally Considered".

4.2.8. Local Autonomy

Another interesting point of comparison between Al Salamat and Shaubat Al Wuttah is the involvement of residents in the design stages of single-family houses of Shaubat Al Wuttah. For one week, Al Ain municipality and Al Dar consultancy firm held meetings with all prospective owners to discuss the design of the housing models (Figure 12). Most owners were satisfied with the designs of their houses after undertaking some suggested modifications. Nonetheless, there is no consideration for residents' involvement in maintaining or managing local resources and projects such as the park and the sports field.

Local production of food is not mentioned in the design documents and it would be left to residents to cultivate some food if they wish in the small green areas inside their housing plots. Renewable energy resources have not been considered in the design either. The results of the assessment of these design elements indicate that the Local Autonomy principle is "Averagely Considered".

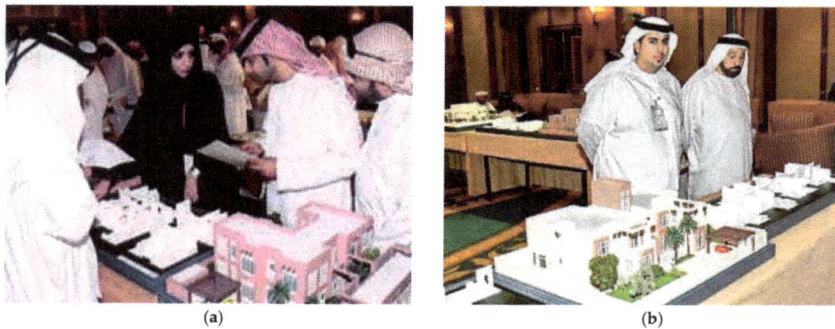

Figure 12. (a,b) Discussing Housing Designs with Residents. (Source: Al Ain Municipality).

4.2.9. Environmental Quality

As in the case of Al Salamat, environmental quality will suffer, though to a marginally lesser degree, from the heavy reliance on private cars. Locations of services within the neighborhood do not encourage residents to walk or cycle on attractive pedestrian pathways. On the contrary, insufficient services and activities mean that residents are more likely to use their cars for lengthy trips. Bus stops and stations, if properly located, might encourage people to use public transportation instead of their own cars. The location of Shaubat Al Wuttah near an industrial area (a cement factory) poses some environmental hazards but little effort to mitigate the effect of such a hazard is noted in the design. Some other measures have been introduced to enhance environmental quality, including highly energy-efficient street lighting. The results of the assessment of the design elements of the Environmental Quality principles indicate that it is "Significantly Considered".

4.2.10. Community Safety

Street design in Shaubat Al Wuttah is generally safe as it considers safety zones and buffers to isolate pedestrian walkways from vehicular movement (Figure 10b). Visual surveillance in the public realm is expected to be better than in Al Salamat in light of these design measures but this would be affected by the lack of mixed use and inappropriate catchment areas, which will discourage residents' presence in the streets. High fences are another obstacle to proper visual surveillance. Moreover, speed control signs and street humps within the neighborhood are expected to contribute to road safety. Facilities and services are designed to be accessible to children, the disabled, and older people, with ramps, signs, and other measures being considered. The results of the assessment of these design elements make this principle "Averagely Considered".

4.2.11. Privacy

Unlike the conventional design in Al Salamat, the design of Shaubat Al Wuttah lacks wide open spaces within the housing plots. Other private spaces such as roof gardens and terraces are not considered in the design (Figures 9 and 11). Based on the results of the assessment of these design elements, the Privacy principle can be considered "Marginally Considered".

4.2.12. Imageability/Sense of Place/Identity

The unique urban design of Shaubat Al Wuttah might help create some distinguishable features and sets of activities that might take place in the small gardens. The architectural design of the houses reflects the historical and traditional architectural character of the UAE. However, consistency among the three adopted architectural design styles (Islamic, Modern, and Traditional) is questionable. On the other hand, the neighborhood location contains an important natural asset of Hafeet Mountain foothill that has not been considered in the urban design to help create a distinctive identity. The results of the assessment of these design elements make this principle "Averagely Considered".

The overall results of the analysis of the urban design principles of Shaubat Al Wuttah local community are summarized in Table 2.

5. Discussion

The results of the analysis of the two social housing case studies—Al Salamat neighborhood representing the past convectional social housing projects in the UAE; and Shaubat Al Wuttah, representing the present turn towards sustainable-orientated social housing—have revealed positive and negative aspects relevant to urban forms with generally better but still insufficient moves towards sustainable urban forms in the recent design of Shaubat Al Wattah. Table 2 summarizes the challenges facing the realization of sustainable urban forms in the Shaubat Al Wuttah social housing project compared to the conventional neighborhood. It has been found that five principles are still marginally considered and five are averagely considered. Only two principles have been significantly considered in the urban form design. These results highlight the type of change in urban form design and its associated challenge, as seen in the last column in Table 2.

Table 2. Assessment of the Consideration of Sustainable Urban Form Design Principles in Both Al Salamat and Shaubat Al Wuttah.

Principle	Al Salamat, Conventional Neighborhood.	Shaubat Al Wuttah, Claimed as Sustainable Neighborhood.	Type of Change in Urban Form Design (Challenges)
Density/Compactness	Marginally Considered	Marginally Considered	Inadequate
Accessibility	Marginally Considered	Averagely Considered	Better but Insufficient
Choice	Marginally Considered	Averagely Considered	Better but Insufficient
Mobility	Marginally Considered	Significantly Considered	Appropriate
Mixed Use	Marginally Considered	Marginally Considered	Inadequate
Social Mix/Social Capital	Marginally Considered	Marginally Considered	Inadequate
Adaptability/Resilience	Averagely Considered	Marginally Considered	Inadequate
Local Autonomy	Not Considered	Averagely Considered	Better but Insufficient
Environmental Quality	Averagely Considered	Significantly Considered	Appropriate
Community Safety	Averagely Considered	Averagely Considered	Better but Insufficient
Privacy	Significantly Considered	Marginally Considered	Better but Insufficient
Imageability/Sense of Place/Identity	Marginally Considered	Averagely Considered	Better but Insufficient

Accordingly, the four major challenges facing the turn towards sustainable urban design, not only in Shaubat Al Wuttah but in other new sustainability-orientated projects in the UAE, are 'low density', 'lack of mixed use', 'lack of social mix' and 'lack of community participation'. The significance of low density comes from the fact that it adversely affects other principles such as choice, mobility (with public transportation), mixed use, and social mix/social capital. For example, the currently designed gross population density in Shaubat Al Wuttah would not be sufficient to support a feasible public transportation system and a mixed-use design.

Therefore, for the Density/Compactness principle, the aim should be to increase the gross density rate to be, as much as possible, closer to the global gross density of 50 to 60 pph. This could be achieved through introducing multi-story, medium-rise apartment blocks with multiple housing units. Insisting on single-family housing as the sole social housing pattern is not going to help produce sustainable urban forms, even if many other sustainability measures are applied. Multi-family housing can be introduced as another type of social housing for Emirati citizens alongside single-family housing when developing social housing neighborhoods. A real challenge here is going to be the local community's acceptance of this type of housing, but with the recent development of some social housing projects in Al Fujairah and Al Ain it might be expected that an increasing number of Emirati citizens are getting closer to accepting living in apartments within multi-story residential blocks. Once intensified, density will help realize efficient public transport and economic vitality, which will, in turn, encourage diversification of services and utilities with a mixed-use design. Without achieving that, sustainable urban design will be difficult to realize.

The unappealing modes of mobility mainly result from inappropriate catchment areas because it is not enough for the design to provide pleasant and safe pedestrian walkways and cycling lanes to convince residents to leave their own cars behind and walk or cycle. Therefore, the neighborhood urban design should create standard catchment areas so that various services and facilities can be reached by different modes of travel, especially by walking and cycling. Again, without higher densities and more compact urban forms, catchment areas will remain wide and walkability and cycling would not be encouraged.

Choice can be enhanced through providing more options for services and facilities, but this needs to be thought about carefully in terms of the assumed population density that can support such variety. Mobility choice has been significantly improved with the introduction of different sustainable modes of travel within the neighborhood, but low density remains a concern for the economic feasibility of public transport. Therefore, increasing density, besides other regulative measures such as paid parking and positing taxes on fuel, are crucial measures for public transportation to function properly. Also connections to other surrounding neighborhoods should be considered more in neighborhood urban design, as currently the most attention is given to individual neighborhood design without considering its possible relationship to its surrounding urban context.

Absence of social mix/social capital is an obvious shortcoming that can be rectified through diversification of housing types and enhancing the potential for residents to meet via locally organized social activities within the shared urban spaces and community facilities. Adaptability requires more attention on both the building and urban space levels. The development of pre-designed, limitedly expandable house models is not sufficient for achieving adaptability and resilience. Rather, there is a need for innovative design solutions that allow for various scenarios for expansion and adaptability according to the changing needs of the community. Residents should enjoy the ability to genuinely change or extend the residential spaces inside their homes, while urban spaces should be designed to accommodate various changing social and economic activities. Besides the provision of all required services and facilities, achieving local autonomy requires the involvement of community members in managing their neighborhood resources, including food production. Environmental quality requires more reliance on sustainable modes of travel such as walking, cycling, and public transport. To encourage people to walk or cycle, sufficient services and amenities should be provided, including working places, within a walkable distance from homes' front doors in a mixed-use design pattern. Utilization of renewable energy resources and energy efficiency measures should be considered if a better environmental quality is to be achieved.

Community safety requires more urban surveillance through urging people to walk and cycle, and, as mentioned above, through developing a more dense and mixed-use residential built environment. Meanwhile, privacy can be enhanced by introducing more private spaces as wider green areas within the housing plots besides green roof and terraces. Finally, enhancing imageability, sense of place, and urban identity can be realized through developing architectural style(s) that are linked to the culture of the local community and through paying more attention to integrating natural assets within the urban design. More citizen involvement would lead to a greater sense of belonging and help produce a unique character for the local community shaped by the laypeople and not imposed on them in a form drawing-board-prepared design styles.

6. Conclusions

Federal and local social housing programs in the UAE are currently witnessing an ambitious and challenging turn from conventional neighborhood design paradigm, with its sprawl, low density and separate zoning, into design patterns that aim to accomplish sustainable urban form principles. This challenging turn has met with some success but it seems that there are still some major obstacles in the way that the new design failed to properly consider. These are 'low population density', 'lack of mixed use', 'lack of social mix' and 'lack of community participation in the decision making and management of their neighborhoods'.

Higher density, design for mixed use, and social mix with genuine community participation would automatically enhance energy efficiency and cater for a variety of means of mobility in addition to widening the range of services and utilities choices for local residents. However, it should be acknowledged here that, as mentioned by the Smart Growth Network [17], local communities have different needs; therefore, the community residents are the ones who will emphasize some sustainable urban forms principles over others. Rectifying these obstacles and rethinking design concepts is by no means an easy task, but this seems to be the most plausible scenario for future actions.

As there are a limited number of social housing designs that adopt the sustainability agenda in the UAE, it is believed that the analysis of further design models would enrich the argument raised by this research. Nonetheless, it is believed that the results of this investigation would pave the way for a futuristic urban design scenario that, if adopted, will lead to a more sustainable urban form of social housing in the UAE and maybe in other Gulf Cooperation Council (GCC), Arab, and Middle Eastern countries that share many environmental, social, and economic circumstances and for which the UAE is a role model in terms of sustainable policies and actions.

Acknowledgments: The author would like to thank the United Arab Emirates University for funding this research through its Center-based Funding Program-Grant No. 31R104. Also, the author appreciates the effort and dedication of Architectural Engineering Master students in collecting data for this research.

Conflicts of Interest: The author declares no conflict of interest.

References

1. Burton, E.; Jenks, M.; Williams, K. *Achieving Sustainable Urban Form*; Routledge: London, UK, 2013.
2. American Society of Landscape Architects. Professional Practice: Sustainable Urban Development. Available online: https://www.asla.org/sustainableurbandevelopment.aspx (accessed on 13 February 2017).
3. Handy, S.; Niemeier, D. Measuring accessibility: An exploration of issues and alternatives. *Environ. Plan. A* **1997**, *29*, 1175–1194. [CrossRef]
4. Neuman, M. The Compact City Fallacy. *J. Plan. Educ. Res.* **2005**, *25*, 11–26. [CrossRef]
5. Frey, H. *Designing the City: Towards a More Sustainable Urban Form*; Spon Press: London, UK, 1999.
6. Allen, A. Sustainable Cities or Sustainable Urbanisation? Palette, UCL's Journal of Sustainable Cities. Available online: http://discovery.ucl.ac.uk/1353511/ (accessed on 10 May 2017).

7. The Emirates Center for Strategic Studies and Research. *Governmental Housing Strategies in the GCC and Some European Countries*; The Emirates Center for Strategic Studies and Research: Abu Dhabi, UAE, 2009.

8. Unified Housing Portal. Available online: http://www.iskan.gov.ae/Home.aspx (accessed on 13 July 2017).

9. Ahmed, K.G. Urban social sustainability: A study of the Emirati local communities in Al Ain. *J. Urban. Int. Res. Placemaking Urban Sustain.* **2012**, *5*, 41–66. [CrossRef]

10. Jenks, M.; Burgess, R. (Eds.) *Compact Cities: Sustainable Urban Forms for Developing Countries*; Spon Press: London, UK, 2000.

11. Jabareen, Y. Sustainable Urban Forms: Their Typologies, Models, and Concepts. *J. Plan. Educ. Res.* **2006**, *26*, 38–52. [CrossRef]

12. Jones, C.; MacDonald, C. Sustainable Urban Form and Real Estate Markets. In Proceedings of the Annual European Real Estate Conference, Milan, Italy, 2–5 June 2004.

13. Coppolaa, P.; Papab, E.; Angielloc, G.; Carpentieric, G. Urban Form and Sustainability: The Case Study of Rome. *Procedia Soc. Behav. Sci.* **2014**, *160*, 557–566. [CrossRef]

14. Dempsey, N.; Brown, C.; Bramley, G. The key to sustainable urban development in UK cities? The influence of density on social sustainability. *Prog. Plan.* **2012**, *77*, 89–141. [CrossRef]

15. Barton, H. (Ed.) *Sustainable Communities*; Earthscan Publications Ltd.: Cambridge, UK, 2000.

16. CNU (Congress for New Urbanism). The Charter of the New Urbanism: The Neighborhood, the District, and the Corridor. Available online: https://www.cnu.org/who-we-are/charter-new-urbanism (accessed on 11 July 2017).

17. Smart Growth Network. Smart Growth Principles. Available online: http://smartgrowth.org/smart-growth-principles/ (accessed on 11 July 2017).

18. Sustainable Cities Institute. Traditional Neighborhood Development (TND). Available online: http://www.sustainablecitiesinstitute.org/topics/land-use-and-planning/traditional-neighborhood-development-(tnd) (accessed on 12 July 2017).

19. Transit Oriented Development Institute (2017) Transit Oriented Development 10 Principles. Available online: http://www.tod.org/placemaking/principles.html (accessed on 10 July 2017).

20. UN Habitat. A New Strategy of Sustainable Neighborhood Planning: Five Principles, Discussion Note 3 Urban Planning. May 2014. Available online: https://unhabitat.org/Wp-content/uploads/2014/05/5-Principles_web.pdf (accessed on 17 July 2017).

21. Center for Liveable Cities. *10 Principles for Liveable High-Density Cities*; Centre for Liveable Cities and Urban Land Institute: Singapore, 2013.

22. Burke, M.; Brown, A.L. Distances people walk for transport. *Road Transp. Res.* **2007**, *16*, 16–29.

23. Biddulph, M. *Introduction to Residential Layout*; Butterworth-Heinemann: Oxford, UK, 2007.

24. ADUPC (Abu Dhabi Urban Planning Council). *Plan Al Ain 2030 Urban Structure Framework Plan*; ADUPC: Abu Dhabi, UAE, 2009.

sustainability

MDPI

Article

Improving Thermal Comfort of Low-Income Housing in Thailand through Passive Design Strategies

Nafisa Bhikhoo [1,*]**, Arman Hashemi** [2,3] **and Heather Cruickshank** [3]

[1] Department of Engineering, University of Cambridge, Cambridge CB2 1PZ, UK

[2] School of Environment and Technology, University of Brighton, Brighton BN2 4GJ, UK; a.hashemi@brighton.ac.uk

[3] Centre for Sustainable Development, Department of Engineering, University of Cambridge, Cambridge CB2 1PZ, UK; hjcruickshank@gmail.com

* Correspondence: nafisabhikhoo@gmail.com; Tel.: +27-796-504-785

Received: 30 June 2017; Accepted: 10 August 2017; Published: 15 August 2017

Abstract: In Thailand, the delivery of adequate low-income housing has historically been overshadowed by politics with cost and quantity being prioritised over quality, comfort and resilience. In a country that experiences hot and humid temperatures throughout the year, buildings need to be adaptable to the climate to improve the thermal comfort of inhabitants. This research is focused on identifying areas for improving the thermal performance of these housing designs. Firstly, dynamic thermal simulations were run on a baseline model using the adaptive thermal comfort model CIBSE TM52 for assessment. The three criteria defined in CIBSE TM52 were used to assess the frequency and severity of overheating in the buildings. The internal temperature of the apartments was shown to exceed the thermal comfort threshold for these criteria throughout the year. The internal operating daily temperatures of the apartment remain high, ranging from a maximum of 38.5 °C to a minimum of 27.3 °C. Based on these findings, five criteria were selected to be analysed for sensitivity to obtain the key parameters that influence the thermal performance and to suggest possible areas for improvement. The computer software package Integrated Environmental Solutions—Virtual Environment (IES-VE) was used to perform building energy simulations. Once the baseline conditions were identified, the software packages SimLab2.2 and RStudio were used to carry out the sensitivity analysis. These results indicated that roof material and the presence of a balcony have the greatest influence on the system. Incorporating insulation into the roof reduced the mean number of days of overheating by 21.43%. Removing the balcony increased the number of days of overheating by 19.94% due to significant reductions in internal ventilation.

Keywords: thermal comfort; low income housing; Thailand; tropical climates; dynamic thermal simulations; sensitivity analysis

1. Introduction

The consequences of rapid urbanization and the growing disparities in wealth between residents in the developing world have brought the issue surrounding the sustainability of low income housing to the forefront. The correlations between population growth, climate change and energy efficiency in housing in these regions indicate that priorities need to be placed on planned future development [1,2]. The accessibility of affordable housing is limited by the socio-economic status of those who need it [3] and the quality of the current stock of low income housing is characterised by technical inefficiencies and inappropriate design elements thus rendering it inadequate for day to day living. With concerns growing over urban liveability in these regions, priorities need to be placed on planned future development [2]. This involves a shift towards the provision of housing that not only make use of environmentally sensitive construction materials, processes and technologies, but also considers how housing performs under the effects of both internal and external climatic factors [1].

In Thailand, low and middle income housing is provided by the government [4]. However, delivery of adequate housing has historically been overshadowed by politics with cost and quantity being prioritised over quality, comfort and resilience [5]. In a country that experiences hot and humid temperatures throughout the year, buildings need to be adaptable to the climate in order to improve the thermal comfort of inhabitants. Extensive research has been done to address energy and thermal comfort issues in developed countries for domestic and non-domestic building. Less research has been done to evaluate and address overheating and thermal discomfort in low-income tropical housing. This research aims to address this by evaluating and suggesting solutions to improve thermal comfort in low-income housing in Thailand.

In trying to overcome challenges of demand and to optimise land usage, low income housing designs in tropical regions were produced and are continuing to be produced according to western standards [3]. An example of the continued implementation of capital intensive methods of "providing large-scale housing to as many people as possible" [3] in Thailand, is the Baan Ua-Arthorn project in Bangkok. Under the Baan Ua-Arthorn programme, roughly 71% of the houses built were low-rise condominiums. The average construction cost of one of these low income condominiums equates to 8000 THB per m^2 [6]. Due to the low cost nature of these housing estates, the units are characterised by their use of inadequate materials [7], the inferior quality of the design and the construction, and located in hard to reach urban zones [8]. While the Baan Ua-Arthorn housing programme was discontinued and replaced with preferred bottom-up or "community-based development" [8] initiatives, the programme highlights the concern over sustainable housing standards for the poorer sectors of society.

The attributes of housing in Thailand are progressively changing due to advances in the socio-economic situations of individuals and the social aspirations attributed to development [9,10]. Figure 1 shows the trend of increased dependency on mechanical forms of cooling within urban areas of Thailand. The incorporation of architectural specifications for housing which are incompatible with both the prevailing climatic conditions is found to exacerbate issues associated with extreme indoor temperatures and comfort, adequate natural ventilation and low levels of indoor air quality in these dwellings [11]. Rapid urbanization has also played a part in the influence of elevated temperatures in these dwellings. Sprawling urban structure and high building density in urban centres such as Bangkok have been found to exacerbate the urban heat island effect (UHI) by decreasing air velocity and increasing air temperatures of the urban climate [12,13]. This has induced a dependency on mechanical forms of cooling once individuals can afford it [14] and the residential energy consumption in Thailand set to increase more than twofold by 2030 [6]. The construction of housing that can adapt to dominant climatic conditions is a key element of providing appropriately sustainable housing and reducing energy consumption in an urban context [2].

Figure 1. Air conditioning units arranged on low income housing units in Phuket, Thailand.

According to the Köppen Climate Classification [15], the tropics fall within 15° north and south of the Equator and are characterised by annual air temperature above 18 °C (64 °F) and the lack of definitive thermal seasonal changes. The temperature ranges between day and night tend to be greater than those experienced between the summer and the winter months. Three variable climatic zones exist within the tropics itself, based on the distinctions experienced in the temperature and precipitation patterns. These are classified as Type A climates and includes the: wet equatorial climate (Af), the tropical monsoon and trade-wind littoral climate (Am) and the tropical wet–dry climate (Aw).

Thailand falls within the tropical wet–dry climate (Aw) characterised by hot and humid conditions throughout the year [10,16]. Three distinct climatic periods with the hottest temperatures experienced from March to May, the rainy season consisting of elevated levels of relative humidity occurs from June to October and a relatively colder period occurs from November to February [14]. The mean daily temperature ranges 26–36 °C with the average minimum temperature falling to 21 °C in the "winter" months with the annual average temperature reaching 28 °C [10]. The relative humidity remains high throughout the year averaging 74–85% and peaking during the rainy months. Daytime temperatures are found to exceed those temperatures deemed thermally comfortable throughout the year [6].

Traditional steady-state thermal comfort models have been found to disregard how people adapt to their environments by changing the conditions to become more accommodating [17–19]. This is of particular concern in hot and humid tropical climates as the application of these thermal comfort indices has been shown to inadequately predict levels of thermal comfort in these regions [7,10,20,21]. In the tropics, people have adapted to being comfortable at higher temperatures for longer periods of time [21]. In Thailand, field studies have shown that in naturally ventilated buildings, individuals remain a state of reasonable comfort at 28 °C [22] with an upper limit for thermal comfort reaching 31.5 °C [10].

The combination of building physics principles and climate is confirmed as an important factor in low income housing design, with inefficient building performance elevating already heightened indoor temperatures and thus impacting on thermal comfort of the inhabitants of these houses [11,23]. The distinct nature of the Thai climate means that housing design in these regions needs to incorporate strategies that exploit the benefits from the outdoor climate to achieve thermal comfort inside [24]. Passive design strategies have been proposed as an adequate method to achieve optimum indoor environmental conditions in residential buildings and thus reduce energy consumption in numerous tropical regions. The main design consideration is incorporating elements that minimise internal heat gains and maintain thermal comfort of inhabitants during periods of high solar radiation and relative humidity. Various studies have been conducted on the types of passive design features that induce improved thermal comfort in tropical regions [20,25–28].

In tropical climates, sufficient air movement through buildings has the capacity to reduce thermal discomfort. The high outdoor temperatures and elevated levels of relative humidity mean that indoor comfort is promoted through both the number of air exchanges that occur as well as the speed of the air [29] Studies have shown that air movement of up to 1 m/s can reduce internal operating temperatures by 3.5 °C [10,21].

To take advantage of ventilation cooling strategies, the distribution, size and number of openings needs to be maximised. Openings in each room are critical to aid in the airflow. The internal layout should be designed to allow for airflow through principal rooms and from the front to the back of the building [30]. The incorporation of balconies induces the free movement of air into tropical housing designs have been found to accelerate airflow into a dwelling [29]. A balcony acts like a "wind scoop" enhancing the rate of air movement through its opening [29].

Stack ventilation is a common form of heat dissipation that uses physical concepts of air density or the stack-effect to drive warm air upwards. The incorporation of a solar chimney on the roof removes the warm, humid air and entrains cooler air into the internal environment. These design parameters have been found to increase the rate of natural ventilation in areas with limited wind speeds [11].

In tropical regions, nighttime temperatures remain relatively high. This makes night cooling of buildings a challenge. The utilisation of ventilation and heat diffusion techniques can help compensate for this climatic restriction and assist in the reduction of internal operating temperatures at night [6].

Radiant gains from sun exposure and conduction gains through the building envelope account for 80% and 20% of external heat gains in tropical climates respectively [30]. Passive design means incorporating efficient mechanisms of solar protection to reduce direct solar exposure. The strategy is to optimise the effects of shading by orientating openings with overhangs (horizontal shades) to the north or south to maximise shading when the sun is at its peak. This provides optimum shading of windows and walls over the year [6]. Vertical shading elements should be used on the east and west facing openings to obstruct direct sunlight.

The thermal characteristics of materials have a significant influence on the induction of passive cooling in buildings. The optimisation of materials with high thermal resistance (low U-values) means that the building envelope will have greater insulating properties, thereby reducing conduction heat into a dwelling. Thermal insulation in the roof and the walls induces the same effects by reducing the heat gain through the structural elements. Conversely, the insulation will restrict heat loss from the interior space at night creating discomfort [6]. Materials with high surface emissivity can easily absorb and release radiant heat which could also induce discomfort [31]. The utilisation of materials with low thermal storage capacities are optimum for improving the thermal comfort at night as they cool down rapidly [6,23].

The purpose of a sensitivity analysis (SA) is to ascertain how the uncertainty associated with the individual inputs into a model affects the uncertainty of the outputs of the model [32]. Passive design parameters can be screened in order to identify the main factors that have an effect on a desired outcome or system. This involves establishing the critical factors and not the interaction between the factors. The applications of SA techniques are useful for assessing thermal responses of building and data variability [33]. This paper describes the process of using a sensitivity analysis in order to assess an array of passive design parameters and their specific effects on the thermal performance on a case study building. Recommendations on the use of these passive design features can then be made.

The Energy and Low Income Tropical Housing programme is structured to look at elements of sustainable design that alleviate energy dependency in both these areas. This study serves as a component of the continued research into identifying low cost methods of improving thermal comfort through passive design techniques, less energy intensive building materials and adaptive construction techniques in tropical regions. Under the long-term outcomes of the ELITH project, this study aims to analyse elements of the building envelope that influence building performance and thereby make recommendations on viable options to solve the inadequacies. The first objective from this research is to develop a detailed analysis on the performance of key design and material elements of government provided low income housing in order to understand contextual aspects of thermal comfort and cooling. The second objective involves assessing the sensitivity of these elements using representative passive design parameters to understand the use of passive design techniques as a low cost design strategy for more sustainable housing supply. Finally, this research aims to make recommendations based on the adequacy of the design strategies in Naturally Ventilated Buildings (NVB) in consideration of the Thai context.

2. Methodology

The methodological approach to answering the research objectives incorporates the use of energy modelling software to obtain data about the thermal performance of a "typical" housing unit in Thailand. The primary processes involved in this study include:

- Establish a housing design to be used as a baseline example for assessment.
- Identify material compositions and geometric design aspects of the condominium housing models.
- Generate a 3-D model incorporating construction and thermal properties of the condominium housing models in IES VE.
- Carry out building energy simulations for the standard housing design using IES-VE software (Integrated Environmental Solutions, Glasgow, United Kingdom).

- Validate the results of the thermal performance of the baseline model according to adaptive thermal comfort standards.
- Identify passive design techniques in the form of key material and design parameters that influence thermal comfort in low income housing in tropical regions.
- Develop permutations of design parameters using SimLab2.2 Sensitivity Analysis Software (Joint Research Centre – European Commission, Brussels, Belgium).
- Run building energy simulations incorporating of the permutations of the standard baseline housing model that incorporate the various design and material changes in IES VE software.
- Compare the results to those obtained for the baseline model in order to assess variances in the thermal performance of the housing model.
- Identify the parameters that have the most effect on the thermal performance.
- Undertake a sensitivity analysis based on the results of running simulations for each permutation using SimLab2.2 Sensitivity Analysis and RStudio statistical software (Allaire Corporation, Newton, MA, USA).

Dynamic Thermal Simulations (DTS) were conducted in IES VE in order to obtain data about the thermal performance of a "typical" housing unit in Thailand. The virtual environment was established using the daily temperature over a twelve-month period for the Bangkok Metropolis (13.73° N, 100.57° E). This included the Dry Bulb temperature and the Daily Running Mean Temperature from the 1st of January to the 31st December. The model was set up in order to assess the performance of the dwelling under the "worst case scenario" conditions for the south facing case study building. It should be noted that, in areas close to the equator, the sun is high in the sky and east/west orientation may represent the worst case scenario [34].

The Baan Ua-Arthorn housing programme was selected for analysis due to its status as a pioneering governmental approach to deliver one million affordable homes for urban citizens in five years [8]. During the eight years of implementation of this programme, the government delivered a total of 253,164 housing units of which 186,507 were condominiums [5]. Statistics show that an average Thai household consists of four individuals [35], thus each level of the apartment block was reduced to a representative five zone layout of 33 m^2 with four occupants. The bedroom is the only room with an outside facing window, while the kitchen and the living room both contain windows overlooking the internal hallway. The layout of a single level is shown in Figure 2. The hallway is 36 m long with a stairwell at one end connecting each of the levels. Each floor contains six apartments. The hallway contains two windows at one end closest to the stairwell.

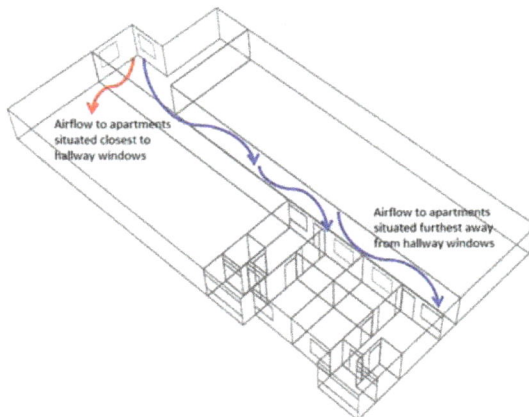

Figure 2. Plan layout of individual apartment with floor area of rooms.

The windows are set at dimensions of 1.5 m × 1.5 m and are situated 1 m above the base. The doors are set at 0.9 m × 2.5 m. Figure 3 shows the basic plan layout of one of the apartments including the dimensions of the windows and doors.

Room Type	Floor Area (m²)
Balcony	5.25
Toilet	3.0
Living Room	12.75
Bedroom	10.5
Kitchen	7.5

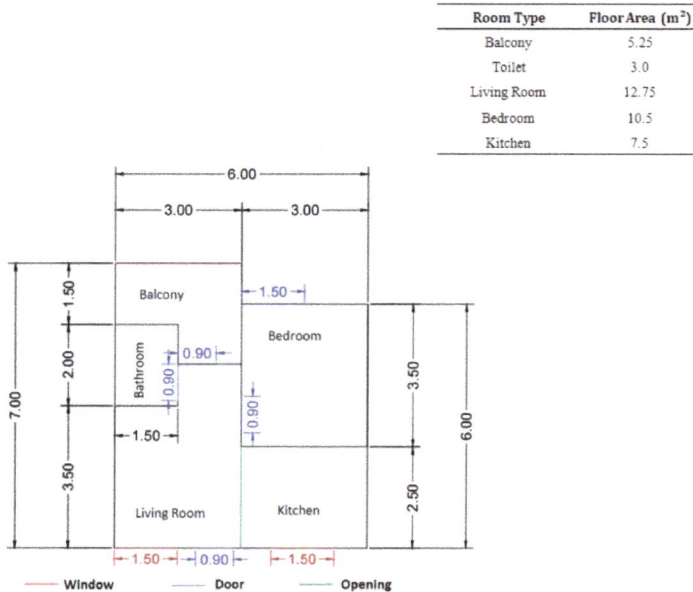

Figure 3. Plan layout of individual apartment with floor area of rooms.

The house windows are all defined as louvre windows (Figure 4) with window openable area at 25%. The openable area for the doors was set to 50%.

Figure 4. Typical example of louvre windows used for housing developments in urban areas of Thailand.

The balcony was modelled as a window that is continuously open at 100%. Figure 5 shows the location of the balcony in proximity to the bedroom. The balustrade of the balcony is composed of clay brick and cement.

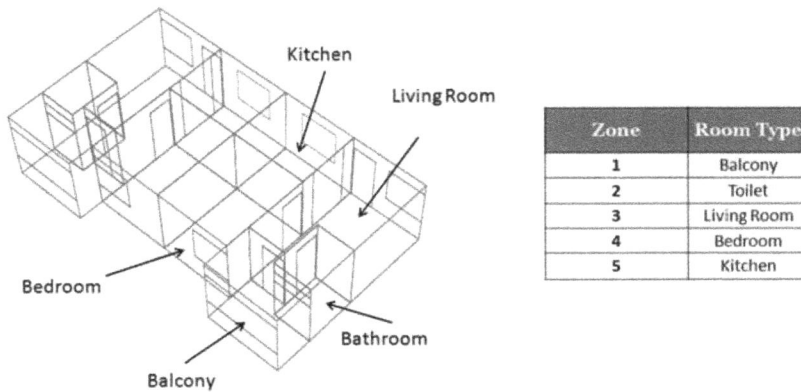

Figure 5. Isometric layout of typical apartment layout with zone classification.

For the basis of this research the buildings are considered as naturally ventilated with no forms of mechanical cooling due to the socio-economic status of the home owners [6]. The number of air exchanges per room was set to 4 ac/h to account for the lower quality standards of the condominiums considering the low income context [8].

According to the Bill of Quantities (BOQ), the standard method of construction for housing under the Baan Ua-Arthorn Housing project included a skeleton composed of reinforced concrete columns and beams for structural stability and infill brickwork walls for the bracing [6]. Table 1 summarises the type and characteristics of the materials used in an individual five-storey apartment block.

Table 1. Material description of typical housing unit [36].

Material	Thermal Conductivity (W/m·K)	Thickness (m)	Surface Emissivity
Roof tiles	6.266	5 mm	0.51
External Walls Clay brick and cement rendering	2.246	200 mm	0.75
Internal Partitions Concrete Block	3.384	100 mm	0.90
Windows glazing	2.7465	6 mm	0.90
Ceiling Gypsum	1.255	90 mm	0.85
Floor Reinforced concrete	3.618	250 mm	0.90

The occupancy profile of the case study building has been specified as fully occupied 20:00–6:00 and on weekends, with working hours spanning 6:00–20:00. The schedule of openings for the baseline model was created on the basis that windows are all open during the day and closed during the night. The simulations for the study are split into two parts, namely Section A: Thermal performance of the baseline model (Table 2); and Section B: The sensitivity analysis for passive design features.

2.1. Section A

Section A details the generation of a 3-D model incorporating construction and thermal properties of the condominium models in IES VE. This includes the validation of the results of the thermal performance of the baseline model according to adaptive thermal comfort standards.

IES Virtual Environment is a software package created by Integrated Environmental Solutions that is used for building energy analysis and sustainable design. IES-VE consists of a range of built-in analysis tools, which facilitate the ease of modelling and analysing the performance of a building either retrospectively or during the design stages of a construction project. The interface makes use of a graphical user interface (GUI) or "black box", which produces graphical results based on a series of user specified inputs. For the purpose of this study the ease of the interface and geometry building, the speed with which results can be produced and the scale of the models needed to be simulated makes it ideal.

The adaptive thermal comfort standard chosen for assessment in this study was CIBSE with the specific guideline CIBSE TM52. Table 2 shows the summary of the conditions chosen for the baseline model. The performance of the baseline model has been studied by reporting the risk of overheating.

Table 2. Summary of conditions specified for baseline model analysis using IES VE.

Specified Condition	Baseline Model Settings
Simulation Period	12 months
Weather Data	Bangkok, Thailand
Occupants	4 occupants with internal gains of 90 W/person/day
Occupancy patterns	20:00–6:00 working days, occupied all other times
Internal Gains	Gas cooking stove 106 W/m², lighting 8 W/m²

Adaptive approach has been used to assess thermal comfort conditions for the baseline model. Thermal comfort in adaptive approach is affected by occupants' behaviours and expectations in naturally ventilated buildings [37]. Based on this method of evaluation, it is proposed that occupants' perception regarding thermal comfort is affected by their thermal circumstances [21]. For typical occupants, CEN standard BS EN 15,251 [38] suggests the following equation to estimate comfortable temperature in naturally ventilated buildings (Equation (1)):

$$Tcomf = 0.33 \, Trm + 18.8 + 3 \text{ (where Trm} > 10 \, ^\circ C), \tag{1}$$

where

Tcomf = the maximum comfortable temperature (°C); and

Trm = the running mean temperature for today weighted with higher influence of recent days [39] (°C).
Trm can be calculated using Equation (2):

$$Trm = [1-\alpha] . \{Ted\text{-}1 + \alpha . \, Ted\text{-}2 + \alpha 2. \, Ted\text{-}3 \ldots .\}, \tag{2}$$

where

Ted-1 = the daily mean external temperature for the previous day (°C);

Ted-2 = the daily mean external temperature for the day before (°C); and so on.

α is a constant; Tuohy et al. [40] suggest to use 0.8 for α.

The CIBSE TM52 guideline assesses performance against three criteria. A zone is classified as overheating if it fails any two of the three criteria [41]. The criteria are defined in terms of ΔT, which is the difference between the actual operative temperature and the maximum acceptable temperature (Table 3). This is rounded to the nearest whole degree.

$$\Delta T = Top - Tmax, \tag{3}$$

Operative temperature (Top) articulates the joint effect of air temperature and mean radiant temperature along with the internal air movement as a single representative figure. For indoor airspeed less than 0.1 m/s, Top is calculated from the following equation [41]:

$$Top = (Ta + Tr)/2 \tag{4}$$

where

Ta = air temperature (°C); and

Tr = mean radiant temperature (°C).

A summary of the overheating assessment criteria is shown in Table 3.

Table 3. Overheating Assessment Criteria.

	Assessment Criteria [1]	Acceptable Deviations
Criterion 1	Percentage of occupied hours during which ΔT ($\Delta T = Top - Tmax$ rounded to the nearest whole degree) is greater than or equal to 1 °C	Up to 3% of occupied hours
Criterion 2	"Daily weighted exceedance" (We) in any one day > 6 °C·h (degree·hours)	0 day
Criterion 3	Maximum temperature level (Tupp) $\Delta T > 4$ °C	0 h

[1] Refer to Abbreviations for more information.

2.2. Section B

The aim of Section B is to assess the sensitivity of 5 passive design parameters which are representative of mitigating/worsening the thermal comfort conditions of the baseline condition. A sensitivity analysis is used to change the parameters in the baseline model to establish the effects that each one has on the system. The effects of various alternations on thermal comfort are carried out using the Simlab and Rsudio programmes. The complete set of baseline conditions and alternative conditions are shown in Table 4.

Table 4. Design options for assessment of sensitivity in condominiums.

Construction	Baseline Condition	Alternative Conditions	
Wall Material	Brick and Cement Rendering	Concrete Block and Rendering	Lightweight Concrete and Rendering
Shading of Windows	Local Shading of Windows	No Local Shading of Windows	
Balcony	Open Balcony	No Balcony	
Window Openable Area	25%	50%	75%
Roof	tiles	tiles with 50 mm insulation	

Based on the results from Section A, a sensitivity analysis (SA) is carried out using two phases, namely screening experiments and optimisation [42]. Firstly, a set of 60 permutations of design parameters is developed using SimLab2.2 Sensitivity Analysis Software in a screening process to establishing the critical factors and not the interaction between the factors. A sequence of results is generated using the Morris Method. This method was selected for this study as it was designed for the screening of a large number of input factors with the outputs incorporating only "elementary effects" [43] of the inputs i.e., those factors with a profound effect on the output and those with a minimal effect. The number of permutations is calculated by the formula

$$r \times (k + 1), \tag{5}$$

where r is the number of levels and k the number of independent input factors [44].

The first step is to convert the factors and levels into a set of permutations for the energy modelling purposes. In the sample generations phase the parameters were identified as a set of discrete variables with a value of 0, 1 or 2 depending on the alternative condition under consideration. The variable passive design strategy parameters (factors) to be assessed for this research are external wall material, the presence of shading devices over the windows, the presence of a balcony, the openable area of the apartment windows and the incorporation of insulation into the roof. The levels for each of these variable design parameters are shown in Table 5.

These variables are then passed through the pre-processor phase where SimLab converts the independent factors and levels into various permutations. The number of permutations that are carried out is dependent on the number of executions selected by the user. For the purpose of this study, the maximum number of executions of 60 was selected to obtain a higher level of accuracy. The permutations are then converted into an experimental matrix under the model execution phase. The complete process undertaken for analysing the incorporation of this data into the energy model represented graphically in Figure 6.

Table 5. Experimental factors and levels.

Criterion	Level Value at 0	Level Value at 1	Level Value at 2
External Wall material (WM)	Baked brick	Lightweight concrete	-
Shading of Windows (SW)	shading	No shading	-
Balcony (BA)	Balcony	No Balcony	-
Window Openable Area (WA)	25%	50%	75%
Roof Material (RM)	No insulation	insulation	-

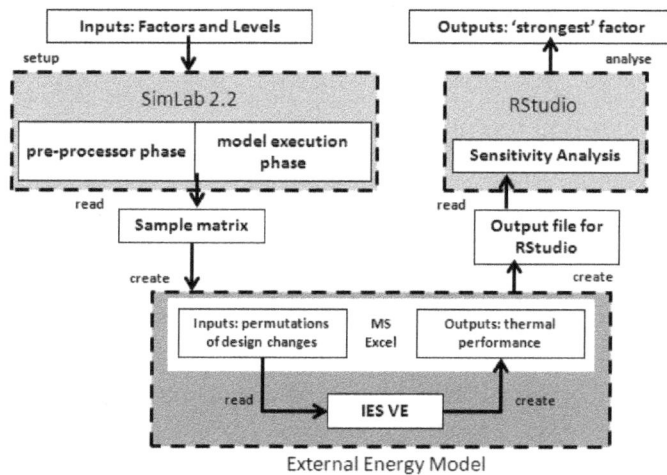

Figure 6. Graphical representation of SA strategy.

The baseline energy model is run for different scenarios, incorporating each of the sixty permutations. Once the simulations were completed in IES Virtual Environment, the optimization phase is carried out. Optimisation refers to the assessment of factor interaction and a system's variance from a statistics perspective. RStudio is used to carry out the optimisation analysis. R is an open source programming language for the statistical analysis. The programme enlists the use of "command-line scripting" for the creation of functions for statistical modelling. The results pertaining to the thermal performance of the model are fed into RStudio. The outputs from RStudio incorporate the sensitivities for each of the independent factors as well as the sensitivities of the factors interacted as pairs. The sensitivities are calculated based on the variances in the values of the three CIBSE TM52 criteria between the baseline model from Section A and each simulated permutation for the living room, bedroom and kitchen. A linear statistical analysis function is run for these living zones for each of the criterion. The R model defines discrete variables in the form of factors by accessing the value of the variable in a column of data to obtain the interaction between factors and can be calculated using Equation (6) below:

$$\text{LR60.model} = \text{lm}(c2 \sim \text{WM} + \text{SW} + \text{BA} + \text{WA} + \text{RM}), \tag{6}$$

where $c2$ refers to the column of data adhering to the CIBSE TM52 data, i.e., $c2$ is Daily Weighted Exceedance. WM is Wall Material, SW is Shading of Windows, BA is the presence of a Balcony, WA is the Window Openable Area and RM is Roof Material. The final output is a graphical representation of the permutations.

3. Results and Discussion

The simulations section evaluates the thermal comfort conditions based on the standard material composition of Baan Ua-Arthorn housing.

3.1. Performance of Baseline Model of Case Study Housing Unit

The results of the thermal performance of the baseline model were validated according to adaptive thermal comfort standards.

3.1.1. Hours of Exceedance (He)

In Figure 7, the percentage of hours of exceedance is shown for the living zones on the ground floor and the fourth floor. The initial observation is that the apartments greatly exceed the limiting factor of 3%. The apartments on the fourth floor are shown to have worse thermal performance than those on the ground floor. The worst performing apartment is the edge unit on the top floor (two exposed external walls). The living room in this unit is the worst performing zone with a performance that exceeds the limiting factor by over five times at a value of 16.14%. The bedroom exceeds the limiting factor by over three times with a value of 10.06%.

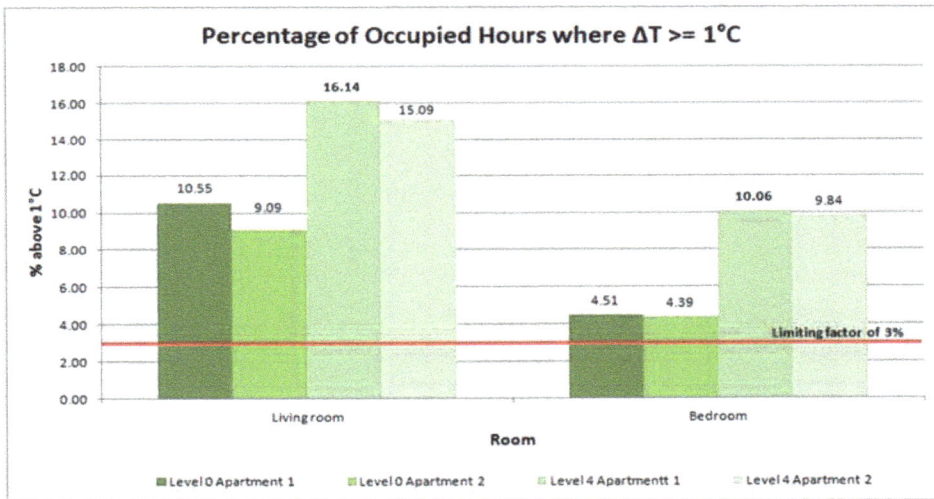

Figure 7. Performance of apartment according to criterion 1.

3.1.2. Daily Weighted Exceedance (We)

This criterion was assessed by counting the number of days in a calendar year where the We exceeds 6 °CHr while that zone was occupied. In compliance with criterion 2, a zone should exceed this value for no days. The results for the baseline case are shown in Figure 8.

As with criterion 1, the apartments are shown to exceed the limits of failure with the corresponding top floor apartment showing the greatest signs of overheating. Within this apartment the living room surpasses 6 °CHr for 115 days and the bedroom surpasses 6 °CHr for 77 days out of 365 days, respectively. This indicates that the zones within the apartment spend a large percentage of time at very high temperatures throughout the year.

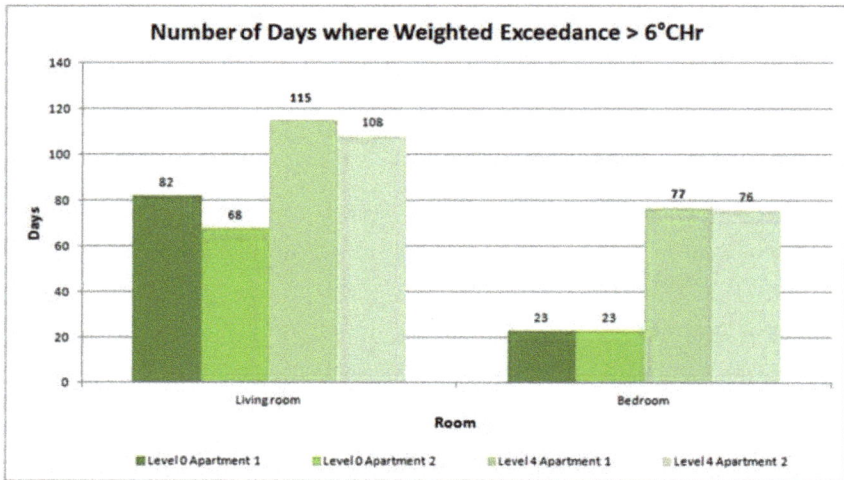

Figure 8. Performance of apartment according to criterion 2.

3.1.3. Daily Weighted Exceedance (We)

The apartments on the lower ground are again found to perform better than those on the top floor. The living room is observed to be the critical zone within the apartments as it fails criterion 3 for three of the four apartments (Figure 9). The differentiation in the performance of the apartments on the lower floor is attributed to the location of the apartments. The unit with two exposed walls (apartment 1) has reduced capacity for providing thermal comfort within the adaptive comfort limits. The living room in apartment 1 on the ground floor and top floor exceed 4 °C by 4 h and 11 h annually, respectively.

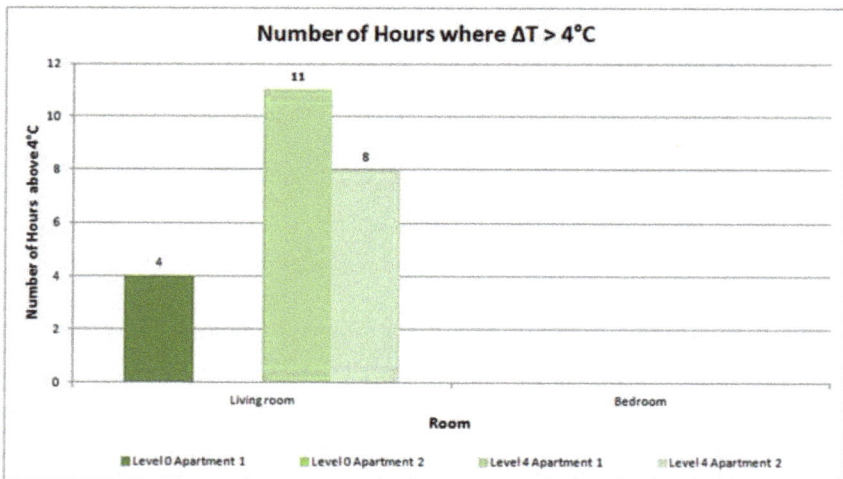

Figure 9. Performance of apartment according to criterion 3.

The zones under consideration within the case study housing unit are found to exceed the acceptable limits of two or more of the CIBSE TM52 criteria. The critical zone of concern is the living room as it incorporates the internal heat gains from the kitchen as these are interleading rooms.

The apartment with the poorer thermal performance was shown to be apartment 1 on the top and ground floors. This is attributed to the material properties of the structural features of the building envelope. This apartment is constructed with two exposed external walls allowing for a higher rate of heat transfer.

In conjunction with the location of the apartments on a level, the height of the condominium influences the thermal performance of the apartments. The building is subjected to effects from "buoyancy-driven air movement" [6]. Hot air from the lower levels rises up through the building and with no means of escaping the living zones, accumulates on the top levels. Combining this with the effects from the building envelope corresponds to the inadequate thermal performance of apartment 1 on level four for all three criteria.

In terms of criterion 3, the bedrooms in each of the apartments do not show exceedance of 4 °C over the year. This can be attributed to the classification of the bedroom as a "night-zone" [30] which means it is only occupied at night. These criteria are assessed based on when the zone is occupied. This means that the external night time temperature drop below a certain point whereby the addition of internal gains from people is not significant enough to raise the temperature above Tupp. In comparison, the living room is either partially or fully occupied at all times. This incorporates those periods where external daytime temperatures reach their maximum.

While these results show that this housing model far exceeds what is deemed acceptable for TM52, it is important to note that TM52 is designed as a tool for mainly assessing overheating in summer in Europe and the UK. Thus, its application to tropical climates tends to underestimate the amount of time spent at high temperatures (which in these regions is most of the day). This is particularly significant for the application of criterion 2. While this level of severity of overheating may be more unacceptable in temperate zones, inhabitants in Thailand are less critical of these conditions. These observations also correlate with those made by Eyre [23] for low income housing in Tanzania and should be incorporated into continued research into establishing adequate thermal comfort criterion for tropical regions [18].

3.2. Summary of Findings

3.2.1. Diurnal Temperature Fluctuation

The 24-h temperature profiles of the living room, the bedroom and the kitchen for the hottest day of the year (29 April) are shown in Figures 10–12, respectively. The variation of room temperature with time shows a low diurnal temperature swing with the internal temperature patterns correlate to the external temperature changes. The internal operating temperatures remain relatively high throughout the day and night, fluctuating between the maximum acceptable temperature and the upper limit for overheating. The external night temperatures do not drop significantly enough to induce rapid cooling of the indoor environment.

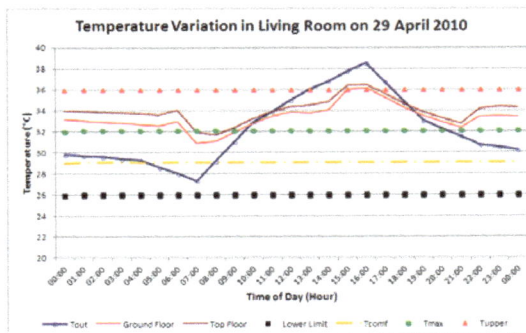

Figure 10. Diurnal temperature variation of living room.

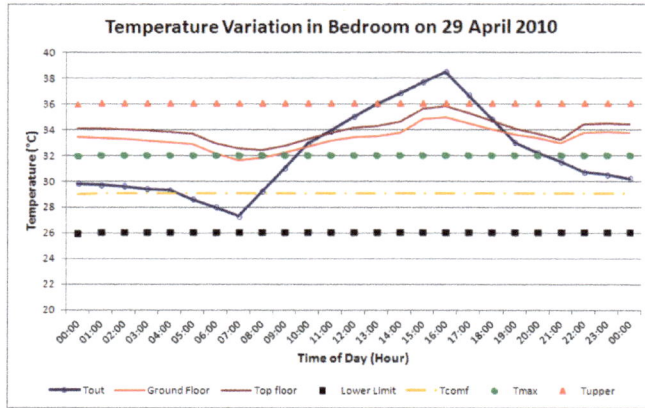

Figure 11. Diurnal temperature variation of bedroom.

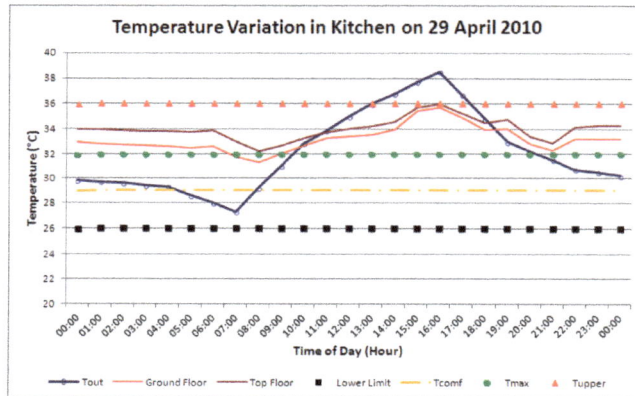

Figure 12. Diurnal temperature variation of kitchen.

3.2.2. Influence of Building Envelope on Thermal Performance

The thermal mass has a significant influence on the cyclical nature of the temperature changes within the apartment units. As the external air temperature rises, the external walls and floor slab will absorb and store the heat. Once the external temperatures start to drop (16:00) the heat within these materials rises to the surface and is released into the internal environment. This elevates the internal night temperatures of the living zones and the indoor temperature starts to drop off at the same time, however at a much slower rate due to thermal storage in the indoor materials. This process is represented in Figure 13, where the fluctuations in the conduction gains of the external walls are influenced by changes in the outdoor temperature. In this case the external walls refers to the impact of all indoor thermal mass (floor, ceiling, internal walls and external walls) releasing stored heat, thus keeping indoor temperatures high. The lack of lag time between peak outdoor and peak indoor temperatures is attributed to the high ventilation rate during the day when the windows are open. This has the resultant effect of moderating the operating temperatures of the living zones. The critical issue is that this effect keeps the operating temperatures at high levels throughout the day, inhibiting sufficient cooling to occur.

Thermal Performance of the External Walls

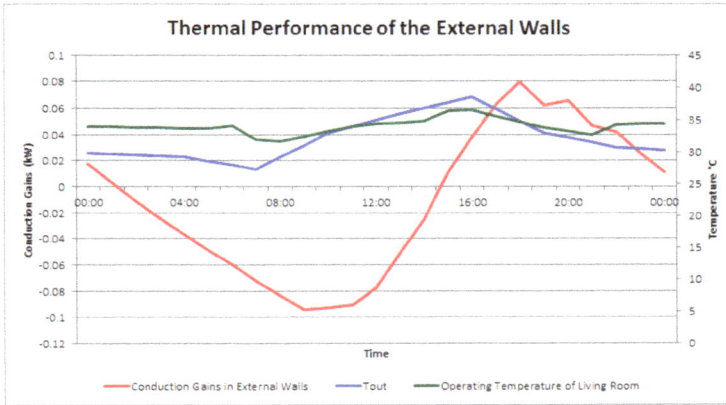

Figure 13. Thermal storage effect of external walls.

In this study, the windows were assumed to be closed at night due to security and social reasons. This limits the amount of airflow in the apartment at night, particularly in the bedroom which has only one window. With insufficient mechanisms to abate excess heat that is released into the zone at night, the operating temperatures of the apartment remains elevated. Figure 14 shows that about a 2 °C reduction in the operating temperature is induced in the bedroom if the window remains open at night and airflow is improved.

Influence of Night Cooling on Thermal Performance

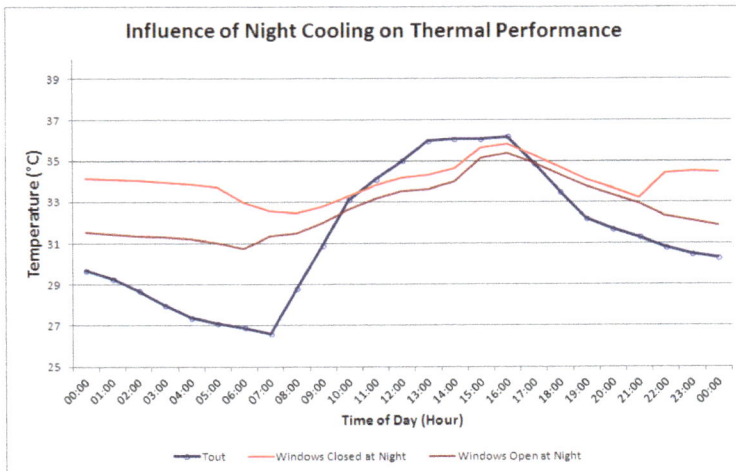

Figure 14. Improvement of ventilation with night cooling.

3.2.3. Influence of Natural Ventilation on Thermal Performance

The high internal operating temperatures are a result of both convection and radiation heat which build up over the day. Without any form of mechanical cooling, natural air exchanges are responsible for the removal of this heat; however, the current construction of the building and each apartment has a significant influence on the ventilation. While the narrow layout may aid in the circulation of air, the number and type of openings, the layout of the rooms and the restrictions of adjacent apartments means that ventilation between rooms is highly restricted. Figure 15 shows the quantity of airflow that enters into each zone. The value Wc refers to the minimum wind speed that is needed to ensure indoor comfort is maintained [10]. The daytime flow rate ranges from 0.11 m/s to 0.38 m/s. The windows remain closed at night which accounts for this rate dropping to zero overnight. The maximum airflow rate in the living room and kitchen is 0.81 m/s and 2 m/s respectively. The high airflow rate in the kitchen is attributed to the presence of the louvre window and the doorway leading into the living room. The window overlooks the hallway and allows for more penetration of airflow through the openings. To achieve a comfortable indoor environment, natural ventilation should provide an indoor air velocity of 0.4 m/s.

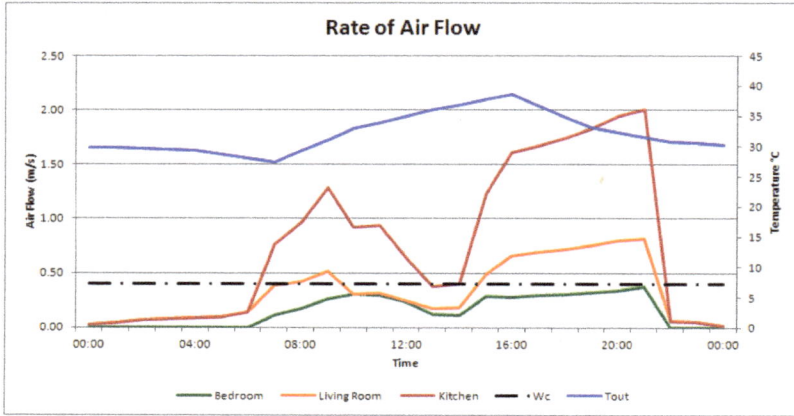

Figure 15. Rate of airflow through the apartment on 29 April 2010.

Essentially, the amount of cross ventilation that can occur through a single unit is highly restricted by design elements and local climatic conditions. The heat builds up and with no method of removal stagnates to increase the operating temperature as well as the discomfort of the internal environment.

3.2.4. Influence of Roof on Thermal Performance

The analysis of the progression of the operating temperature change over the 24 h showed that the roof is subject to a significant temperature change over the course of the day. The temperature change in the roof is seen to begin at 9:00 as the external temperature rises and the solar radiation increases (Figure 16). The temperature of the apartments is seen to be about 5 °C higher than in those on the ground floor at this time. By 14:00, the roof reaches its highest temperature.

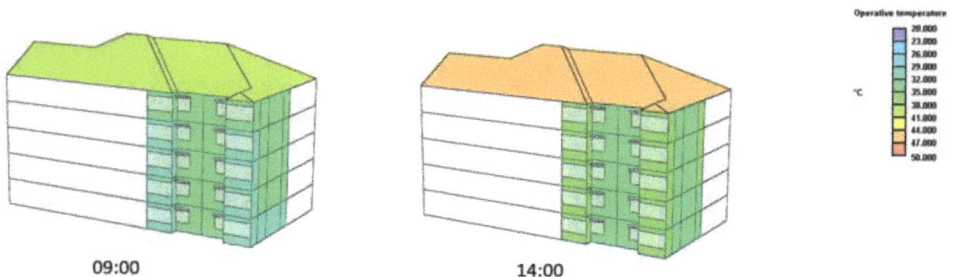

Figure 16. Progression of temperature change in the roof.

The corresponding conduction gains in the roof over 24 h are shown in Figure 17. The conduction values range from a minimum of 1.98 kW at 7:00 to a maximum value of 21.86 kW at 12:00. This corresponds to the increase in direct solar exposure over the day. The negative gains during the night are associated with reversal in the direction of heat transmission, i.e., the roof temperature is higher than the external temperature.

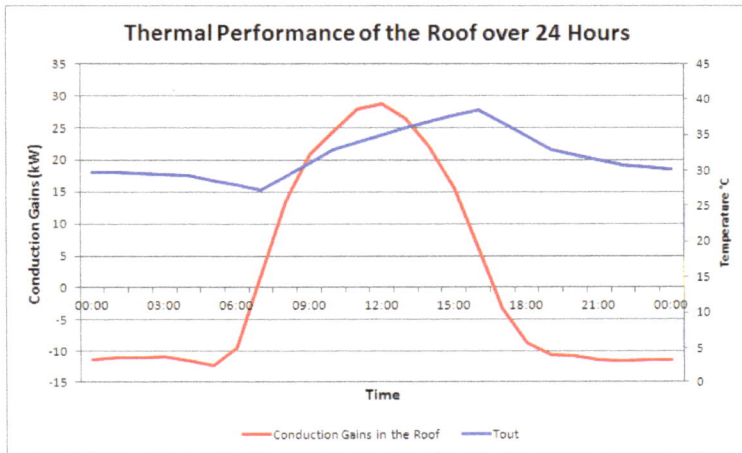

Figure 17. Severity of conduction gains in the roof on 29 April 2010.

Various DTS studies that have been carried out on houses in tropical regions have shown that the roof is a key area of concern in terms of thermal performance [23,24,30,45,46]. The roof is continually exposed to high levels of solar radiation and materials used in roof construction tend to have low thermal storage and low thermal resistance properties. This means that a building remains vulnerable to high levels of heat transmission occurring through the roof. In the case study building, there is a significant difference in operating temperatures between the apartments on the upper level and those on the ground floor. This is partly due to the stack effect of air; however, this can also be attributed to the high magnitude and the rapid transmittance of heat energy through the roof.

3.3. Section B: Summary of Results

The distribution of the data set for each of the parameters is presented in box plot format. The important values include the median or central tendency measurement, the maximum and minimum number of days on overheating. The most important observations that can be made from these results include the distribution of the number of days of overheating that is experienced based on the parameter and its respective level. A change in a parameter that shows a smaller variability in the data set indicates a closer correlation between the individual values in that set (the mean is more representative of the data set). Although the range and the interquartile range are important to show the spread of the data, these points are determined from only two points in the entire data set. Thus from a statistics point of view, the values of the mean and the median are more adequate indications of the sensitivity of the parameter in the performance of the system.

3.3.1. Roof and Wall Material

The incorporation of 50 mm of insulation in the roof is observed to have the greatest change in thermal performance within the apartment (Figure 18). The mean number of days of overheating is reduced to 98.56 when the wall material is changed to level 1, compared to 124 days at level 0.

Insulation reduces the amount of heat gain that can enter into the space between the roof and the ceiling. This means that there is a restriction on the flow of heat in the day and at night. The maximum and minimum number of days of overheating was reduced from 185 to 125 and 115 to 30 days, respectively. The spread of data points is greatly improved by incorporating insulation. This means that less variance is seen in the effects of overheating.

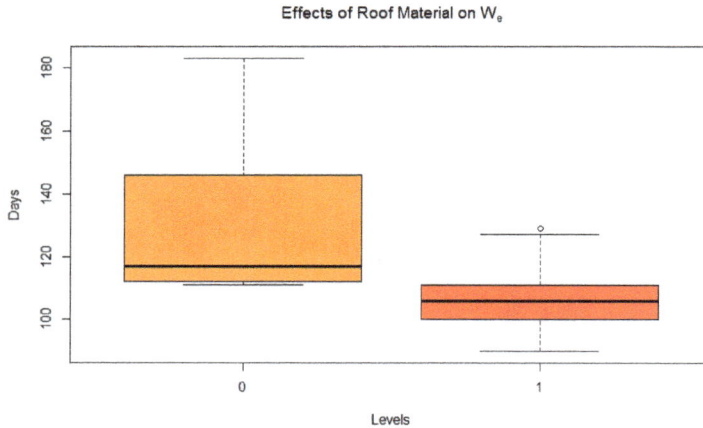

Figure 18. Individual sensitivity: roof material.

The change of the wall material from brick (level 0) to lightweight concrete (level 1) is seen to be less critical than changes in the other parameters. Figure 19 shows the reason for this is that the results of the two data sets overlap significantly. A change in wall material resulted in a reduction of overheating to a mean of 123.97 days, with the median value changing from 170 to 160 days. Although the maximum and minimum number of days of overheating was reduced from 175 to 142 and from 115 to 30 days, respectively, the value of the median is a better indication of the typical number of days of overheating. The outlying value represents a significant irregularity in the distribution which falls outside 1.5 times the upper quartile rage.

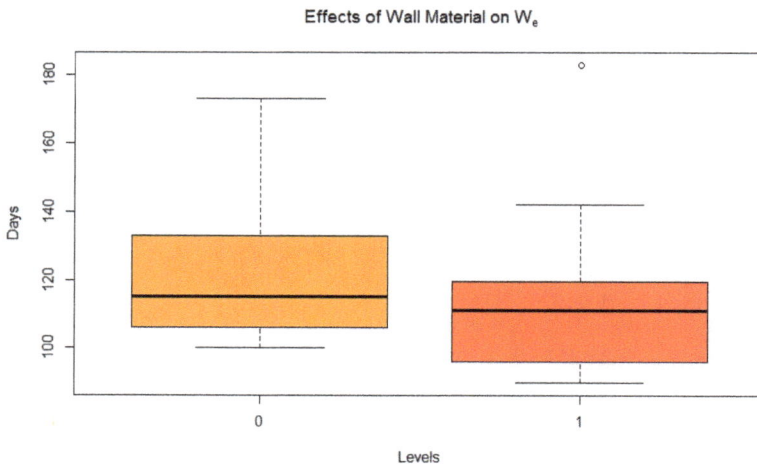

Figure 19. Individual sensitivity: wall material.

The total sensitivity effects of the combinations of uncertainty in the distributions of both wall material and roof material are shown in Figure 20. The levels on the x-axis show the effects of changing first the wall material (1, 0), the roof material (0, 1) and then both wall material and roof material (1, 1). The combined incorporation of 50 mm insulation and the change of wall material to lightweight concrete reduced the mean number of days of overheating to 101.59 with the maximum reaching 130 days compared to 172 days of the baseline condition. The effects in the combination of the sensitivities can be seen on the graph where the median value has decreased from 135 days at the baseline level to 100 days with both parameters set to level 1. The elementary effects of the incorporation of roof material only, however, were shown to reduce the mean number of days of overheating to 98.56. This indicates that although changing the value of these parameters both to level 1 would elicit a decrease in the severity of overheating experienced in the apartment, changing the roof material only has a greater effect in reducing the severity of overheating. As a caveat in multistory buildings, roof insulation primarily impacts thermal conditions on the top floors. In terms of this, if the sensitivity analysis related the impact of action to number of apartments impacted, the outputs yield different results.

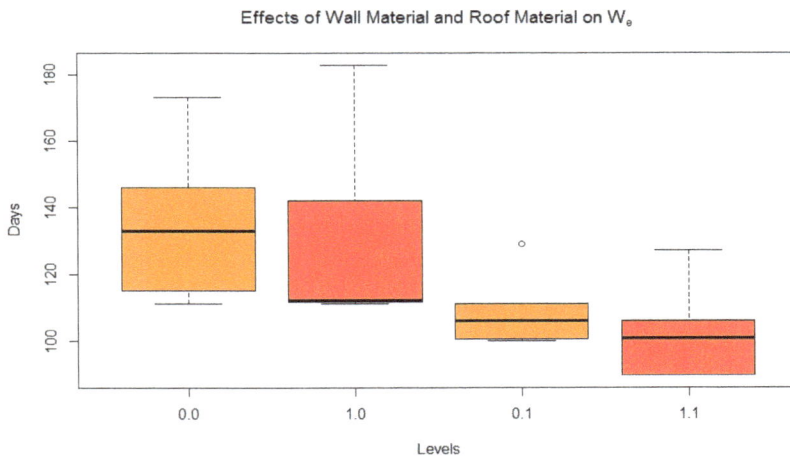

Figure 20. Total sensitivities: wall material and roof material.

3.3.2. Balcony and Window Openable Area

The impact of ventilation on the thermal performance of the building was discussed under summary of findings. The sensitivity of this parameter is essentially assessing for how a decrease in ventilation through the apartment would influence the thermal performance of the design. This parameter was seen to have the second largest effect on the system after the roof material. The graphical representation of these results is shown in Figure 21. At level 1 (the balcony is closed), the maximum number of days of overheating has increased from 118 to 183. The mean was seen to increase to 150.45 days. The already stilted airflow through the living room is further exacerbated once the balcony is removed. This indicates that the temperature of the living zones stays at elevated temperatures for longer periods without any form of ventilation. This is expected in a region with high humidity as well as low minimal air movement. The presence of the balcony has a profound influence on the natural ventilation through the apartment.

Effects of Balcony on W$_e$

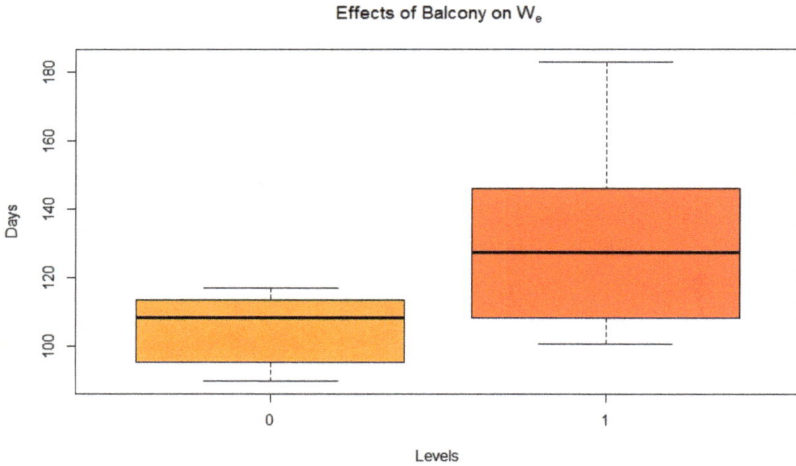

Figure 21. Individual sensitivity: Balcony.

In terms of the openable area of the windows, this is also a measure of the amount of ventilation that is generated in the apartment. This parameter was set at three different levels to assess the extreme variability of the importance of opening size on ventilation and thus the thermal performance of the apartment (Figure 22). The baseline condition for window openable area is 25% (level 0). The mean number of days of overheating at levels 1 and 2 (50% and 75% openable area) were determined to be 116.34 and 110.63 days, respectively. When the openable area is set to 50%, the variability of the data set was increased. The nature of the plot can be attributed to single data points that lie further away from the mean, which is why the median (112 days) is a better representation of this data set. At level 2, the variability of the dataset is reduced and the mean is a more adequate representation of how this parameter influences the performance. Essentially the system is shown to be highly sensitive to a change in this parameter.

Effects of Window Openable Area on W$_e$

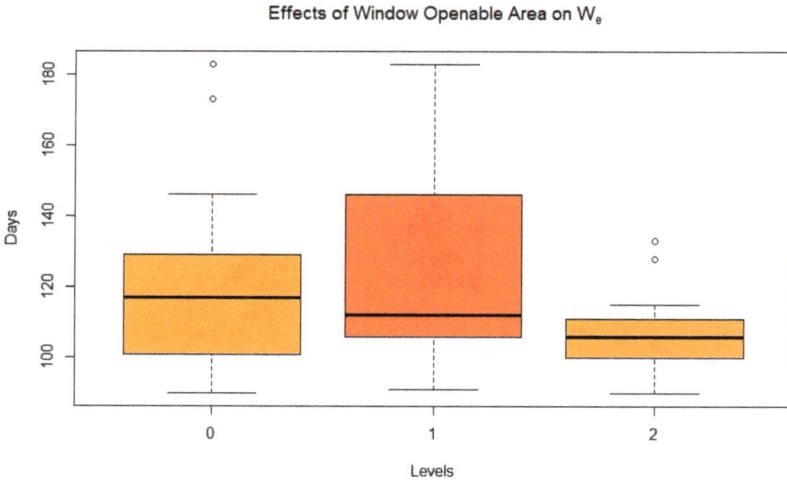

Figure 22. Individual sensitivity of window openable area.

Sustainability **2017**, *9*, 1440

4. Conclusions

In terms of the baseline model, apartment 1 on level four has the worst thermal performance because of its proximity to the roof, the effect of hot air movement into this space and the structural characteristics of the apartment envelope. There is a lack of diurnal temperature variation within the apartments. The internal operating temperatures remain relatively high throughout the day and night, ranging from a maximum of 38.5 °C to a minimum of 27.3 °C. This is not significant enough to influence night cooling. The roof is a key area of concern due to its high U-value, the surface area and the level of exposure to solar gains. This induces a high transmittal of heat into and out of the building. This has a significant effect on the operating temperatures of the apartments on the top floor.

The sensitivity analysis showed that the main effects (effect of a change in one parameter while the others are held steady) of the parameter changes have a greater influence on the performance of the system and are an adequate representation of parameter sensitivity. The system is shown to be most sensitive to a change in the roof material and the incorporation of a balcony into the design. The parameters that are least sensitive to changes in the system include the shading of windows and wall material. The mean number of days of overheating was reduced to 98.56 with the incorporation of 50 mm insulation and increased to 150.45 when the balcony was removed. These effects can be attributed to a reduction of heat gain through the roof when insulation is incorporated but an increase in heat gain due to a lack of ventilation without a balcony opening.

Acknowledgments: This document is an output from a research project "Energy and Low-Income Tropical Housing" co-funded by UK aid from the UK Department for International Development (DFID), the Engineering and Physical Science Research Council (EPSRC) and the Department for Energy and Climate Change (DECC), for the benefit of developing countries. The views expressed are not necessarily those of DFID, EPSRC or DECC.

Author Contributions: Authors equally contributed to the paper.

Conflicts of Interest: The authors declare no conflict of interest.

References

1. Golubchikov, O.; Badyina, A. *Sustainable Housing for Sustainable Cities: A Policy Framework for Developing Countries*; UN-HABITAT: Nairobi, Kenya, 2012; p. 73.
2. Hannula, E.-L. *Going Green: A Handbook of Sustainable Housing Practices in Developing Countries*; UN-HABITAT: Nairobi, Kenya, 2012; p. 124.
3. French, M.A.; Majale, M.; Tipple, G. Affordable Land and Housing in Asia. Available online: https://unhabitat.org/books/affordable-land-and-housing-in-asia-2/ (accessed on 29 July 2015).
4. Kritayanavaj, B. Affordable Housing in Thailand. Available online: https://sites.google.com/site/prcudweb/e-library (accessed on 15 June 2015).
5. Chiarakorn, S.; Rakkwamsuk, P.; Aransiri, K.; Aneksaen, N.; Parnthong, R. Evaluation of greenhouse gas emission from residential buildings in Thailand. In Proceedings of the Asian Conference on Sustainability, Energy & the Environment Official Conference, Osaka, Japan, 12–15 June 2014.
6. Suenderman, T. *Energy Efficiency in Affordable Low-to-Medium-Income Urban Housing of the National Housing Authority of Thailand*; Deutsche Gesellschaft für Internationale Zusammenarbeit (GIZ): Bonn, Germany, 2005.
7. Hwang, R.-L.; Cheng, M.-J.; Lin, T.-P.; Ho, M.-C. Thermal perceptions, general adaptation methods and occupant's idea about the trade-off between thermal comfort and energy saving in hot–humid regions. *Build. Environ.* **2009**, *44*, 1128–1134. [CrossRef]
8. Archer, D. Social capital and participatory slum upgrading in Bangkok, Thailand. University of Cambridge, 2010. Available online: https://www.repository.cam.ac.uk/handle/1810/244821 (accessed on 17 June 2015).
9. Ahmed, K.S. Comfort in urban spaces: Defining the boundaries of outdoor thermal comfort for the tropical urban environments. *Energy Build.* **2003**, *35*, 103–110. [CrossRef]
10. Tantasavasdi, C.; Srebric, J.; Chen, Q. Natural ventilation design for houses in Thailand. *Energy Build.* **2001**, *33*, 815–824. [CrossRef]

11. Santamouris, M.; Pavlou, K.; Synnefa, A.; Niachou, K.; Kolokotsa, D. Recent progress on passive cooling techniques Advanced technological developments to improve survivability levels in low-income households. *Energy Build.* **2007**, *39*, 859–866. [CrossRef]

12. Höppe, P. Improving indoor thermal comfort by changing outdoor conditions. *Energy Build.* **1991**, *16*, 743–747. [CrossRef]

13. Vallati, A.; De Lieto Vollaro, A.; Golasi, I.; Barchiesi, E.; Caranese, C. On the impact of urban micro climate on the energy consumption of buildings. *Energy Procedia* **2015**, *82*, 506–511. [CrossRef]

14. Antarikananda, P.; Douvlou, E.; McCartney, K. Lessons from traditional architecture: Design for a climatic responsive contemporary house in Thailand. In Proceedings of the PLEA2006 23rd Conference on Passive and Low Energy Architecture, Geneva, Switzerland, 6–8 September 2006; pp. 11–43. Available online: http://www.researchgate.net/profile/Kevin_Mccartney2/publication/242123373_Lessons_from_traditional_architecture_Design_for_a_climatic_responsive_contemporary_house_in_Thailand/links/53f3d9fc0cf256ab87b79921.pdf (accessed on 1 August 2015).

15. Arnfield, A.J. Koppen Climate Classification. Available online: http://www.britannica.com/EBchecked/topic/322068/Koppen-climate-classification (accessed on 29 May 2015).

16. Prianto, E.; Depecker, P. Optimization of architectural design elements in tropical humid region with thermal comfort approach. *Energy Build.* **2003**, *35*, 273–280. [CrossRef]

17. Brager, G.S.; de Dear, R.J. Thermal adaptation in the built environment: A literature review. *Energy Build.* **1998**, *27*, 83–96. [CrossRef]

18. Nguyen, A.T.; Singh, M.K.; Reiter, S. An adaptive thermal comfort model for hot humid South-East Asia. *Build. Environ.* **2012**, *56*, 291–300. [CrossRef]

19. Nicol, J.F.; Humphreys, M.A. Adaptive thermal comfort and sustainable thermal standards for buildings. *Energy Build.* **2002**, *34*, 563–572. [CrossRef]

20. Feriadi, H.; Wong, N.H. Thermal comfort for naturally ventilated houses in Indonesia. *Energy Build.* **2004**, *36*, 614–626. [CrossRef]

21. Nicol, F. Adaptive thermal comfort standards in the hot–humid tropics. *Energy Build.* **2004**, *36*, 628–637. [CrossRef]

22. Rangsiraka, P. Thermal Comfort in Bangkok residential buildings, Thailand. In Proceedings of the PLEA2006 23rd Conference on Passive and Low Energy Architecture, Geneva, Switzerland, 6–8 September 2006.

23. Eyre, M.; Hashemi, A.; Cruickshank, H.; Jordan, M. Transition in housing design and thermal comfort in rural Tanzania. In Proceedings of the 5th International Conference on Zero Energy Mass Custom Home (ZEMCH 2016), Kuala Lumpur, Malaysia, 20–23 December 2016.

24. Jayasinghe, M.T.; Attalage, R.A.; Jayawardena, A.I. Thermal comfort in proposed three-storey passive houses for warm humid climates. *Energy Sustain. Dev.* **2002**, *6*, 63–73. [CrossRef]

25. Alvarado, J.L.; Martínez, E. Passive cooling of cement-based roofs in tropical climates. *Energy Build.* **2008**, *40*, 358–364. [CrossRef]

26. Arayela, O. *Sustainable Housing Development Policy for Developing Countries of Africa-Nigeria as a Case Study*; Ural, O., Abrantes, V., Tadeu, A., Eds.; Wide Dreams Projectos Multimedia Lda: Coimbra, Portugal, 2002.

27. Cho, K.-M.; Lee, T.-G.; Han, Y.-H. A Study on Heating Energy Monitoring of a Rural Detached House Applying Passive House Design Components. *J. Korea Inst. Ecol. Archit. Environ.* **2013**, *13*, 39–46. [CrossRef]

28. Jayasinghe, M.T.; Priyanvada, A.K. Thermally comfortable passive houses for tropical uplands. *Energy Sustain. Dev.* **2002**, *6*, 45–54. [CrossRef]

29. Prianto, E.; Depecker, P. Characteristic of airflow as the effect of balcony, opening design and internal division on indoor velocity: A case study of traditional dwelling in urban living quarter in tropical humid region. *Energy Build.* **2002**, *34*, 401–409. [CrossRef]

30. Garde, F.; Boyer, H.; Gatina, J.C. Elaboration of global quality standards for natural and low energycooling in French tropical island buildings. *Energy Build.* **1999**, *34*, 71–83. [CrossRef]

31. The Concrete Centre. MPA-The Concrete Centre. Available online: http://www.concretecentre.com/about_us.aspx (accessed on 3 August 2015).

32. Tarantola, S.; Giglioli, N.; Jesinghaus, J.; Saltelli, A. Can global sensitivity analysis steer the implementation of models for environmental assessments and decision making? *Stoch. Environ. Res. Risk Assess.* **2002**, *16*, 63–76. [CrossRef]

33. Lomas, K.J.; Eppel, H. Sensitivity Analysis techniques for building thermal simulation programs. *Energy Build.* **1992**, *19*, 21–44. [CrossRef]

34. Hashemi, A.; Khatami, N. Effects of Solar Shading on Thermal Comfort in Low-income Tropical Housing. *Energy Procedia* **2017**, *111*, 235–244. [CrossRef]

35. Ministry of Information and Communication Technology. The Gender Statistics Survey. Available online: http://web.nso.go.th/en/survey/gender/gender.htm (accessed on 29 July 2015).

36. Chartered Institution of Building Services Engineers [CIBSE]. *CIBSE GUIDE A: Environmental Design*; CIBSE: London, UK, 2007.

37. The Chartered Institution of Building Services Engineers. *CIBSE Guide a Environmental Design*; The Chartered Institution of Building Services Engineers: London, UK, 2006; Volume 7.

38. British Standards Institution [BSI]. *BS EN 15251: 2007: Indoor Environmental Input Parameters for Design and Assessment of Energy Performance of Buildings Addressing Indoor Air Quality, Thermal Environment, Lighting and Acoustics*; BSI: London, UK, 2007.

39. Nicol, F.; Humphreys, M. Derivation of the adaptive equations for thermal comfort in free-running buildings in European standard EN15251. *Build. Environ.* **2010**, *45*, 11–17. [CrossRef]

40. Tuohy, P.G.; Humphreys, M.A.; Nicol, F.; Rijal, H.B.; Clarke, J.A. Occupant behaviour in naturally ventilated and hybrid buildings. *Am. Soc. Heat. Refrig. Air Cond. Eng. [ASHRAE] Trans.* **2009**, *115*, 16–27.

41. Chartered Institution of Building Services Engineers [CIBSE]. *CIBSE TM52: 2013: The Limits of Thermal Comfort: Avoiding Overheating in European Buildings*; BSI: London, UK, 2013.

42. Telford, J.K.; Uy, M. Optimization by Design of Experiment techniques. In Proceedings of the 2009 I11 Aerospace Conference, Big Sky, MT, USA, 7–14 March 2009; pp. 1–10. Available online: http://i11xplore.i11.org/xpls/abs_all.jsp?arnumber=4839625 (accessed on 20 June 2015).

43. Morris, M.D. Factorial Sampling Plans for Preliminary Computational Experiments. *Technometrics* **1991**, *33*, 167–174. [CrossRef]

44. SimLab 2.2. Simlab 2.2 Reference Manual. SimLab 2.2, 2015. Available online: https://ec.europa.eu/jrc/en/samo/simlab (accessed on 15 May 2015).

45. Hashemi, A.; Cruickshank, H.; Cheshmehzani, A. Improving Thermal Comfort in Low-income Tropical Housing: The Case of Uganda. In Proceedings of the ZEMCH 2015 International Conference, Lecce, Italy, 22–25 September 2015; pp. 22–25.

46. Hashemi, A. Climate Resilient Low-income Tropical Housing. *Energies* **2016**, *9*, 486. [CrossRef]

sustainability

MDPI

Article

Substrate Depth, Vegetation and Irrigation Affect Green Roof Thermal Performance in a Mediterranean Type Climate

Andrea Pianella [1,2,*], Lu Aye [2], Zhengdong Chen [3] and Nicholas S. G. Williams [1]

[1] School of Ecosystem and Forest Sciences, Faculty of Science, The University of Melbourne, 500 Yarra Blvd, Richmond 3070, Australia; nsw@unimelb.edu.au
[2] Renewable Energy and Energy Efficiency Group, Department of Infrastructure Engineering, Melbourne School of Engineering, The University of Melbourne, Melbourne 3010, Australia; lua@unimelb.edu.au
[3] CSIRO Land and Water, CSIRO, Clayton South Victoria 3169, Australia; dong.chen@csiro.au
* Correspondence: andrea.pianella@unimelb.edu.au; Tel.: +61-3-8344-2131

Received: 30 June 2017; Accepted: 10 August 2017; Published: 16 August 2017

Abstract: Green roofs are consistently being used to reduce some of the negative environmental impacts of cities. The increasing interest in extensive green roofs requires refined studies on their design and operation, and on the effects of their relevant parameters on green roof thermal performance. The effects of two design parameters, substrate thickness (ST) and conductivity of dry soil (CDS), and four operating parameters, leaf area index (LAI), leaf reflectivity (LR), stomatal resistance (SR), and moisture content (MC), were investigated using the green roof computer model developed by Sailor in 2008. The computer simulations showed that among the operating parameters, LAI has the largest effects on thermal performance while CDS is a more influential design parameter than ST. Experimental investigations of non-vegetated and sparsely vegetated green roofs in Melbourne were principally used to understand the effect of the substrate and enable better understanding of dominant heat transfer mechanisms involved. Investigated green roofs had three substrate thicknesses (100, 150 and 200 mm), and their performance was compared to a bare conventional roof. In contrast to the computer simulations, the experimental results for summer and winter showed the importance of MC and ST in reducing the substrate temperature and heat flux through the green roof.

Keywords: green roof; substrate; thermal performance; heat flux; parametric analysis; sustainable buildings and cities; energy efficient buildings; climate change mitigation

---

1. Introduction

As part of efforts to reduce air pollutants, greenhouse gases and their carbon footprint [1,2], cities and towns have introduced new technologies and techniques to mitigate some of negative impact of cities on the environment and make cities greener and more sustainable [3–5]. Green roofs, also called vegetated or living roofs, are growing in popularity worldwide and offer a potential solution to the some of the negative environmental impacts of cities [6,7]. Green roofs are engineered ecological systems integrated with the built environment to provide a wide range of ecosystem services, such as air purification [8], social and recreational opportunities [9], mitigation of stormwater runoff [10] and urban heat island effect [11], and they can also provide biodiversity habitat [12]. Green roofs also offer direct energy benefits to buildings and their surrounding areas, such as decreasing building cooling and thermal loads, reducing building energy consumption and, to some extent, mitigation of the urban heat island (UHI) effect [13–16]. Reduction of building energy consumption produced from burning fossil-based fuels helps reduce the emission of greenhouse gases and air pollution [17].

Studies have been conducted all around the world to quantify the extent to which green roofs reduce the heating and cooling loads of commercial and residential buildings. Findings are varied and sometimes in contrast to one another. Common findings are that green roof thermal performance depends on the climate zone, the building materials, the seasonality and the green roof material selection [16,18–23]. For example, in cold climate areas, a thick substrate enhances thermal performance compared to a thin substrate [24]. In contrast, in hot and wet climate areas, a thin 10 cm deep substrate is sufficient to reduce the energy required for cooling the space below [25], while the greatest benefit to green roof thermal performance is offered by a dense and healthy vegetation [26]. For hot and dry climates, appropriate plant selection is essential because green roofs need to be drought tolerant [27].

Because of these wide ranging results, it is not possible to specify one "optimum" green roof build-up (drainage layer, substrate or growing medium and plants) that will maximise green roof thermal benefits in all countries or climate zones. In situ research is therefore necessary to help select substrates and plants for green roofs in various locations [19].

Parametric, also called sensitivity, analyses using existing green roof thermal models can help understand which parameters are most effective in enhancing green roof thermal benefits, and thus maximise green roof thermal performance in different locations. However, this is not sufficient, unless the results of the parametric analysis are validated with field measurements and a comprehensive evaluation.

In the first part of this study, we investigate the effect of two design parameters and four operating parameters using the green roof computer model developed by Sailor [28]. Design parameters refer to properties of green roofs, which do not change throughout the year or the life of the green roof, and are related to the substrate composition and depth. In contrast, operating parameters change during the life of green roofs and influence, and they are influenced by, their vegetation dynamics [29–31].

Subsequently, we present substrate temperatures and heat fluxes of three green roofs in non-vegetated and sparsely vegetated states and the heat flux values of a bare conventional roof to validate the magnitude of Sailor's green roof thermal model outputs. Finally, we discuss the results from computer simulations and experimental green roofs.

2. Materials and Methods

2.1. Computer Simulations

Among the green roof thermal computational models developed and available from the literature [32–35], we selected the model developed by Sailor [28]. The model is a phenomenological model based on the Fast All-Season Soil Strength (FASST) model developed by Frankenstein and Koenig and it is available within EnergyPlus building energy simulation software tool [36].

Sailor's model comprise many variables and parameters, some relevant to the vegetation layer, and others to the substrate layer, called soil in the model. As a result of a review of the literature [19], we selected three vegetation parameters and three soil parameters as the most relevant to a green roof thermal performance for further investigation. These are: (i) Leaf Area Index (LAI); (ii) Leaf reflectivity (LR); (iii) Minimum stomatal resistance (SR); (iv) Substrate thickness (ST); (v) Conductivity of dry soil (CDS); and (vi) Saturation volumetric moisture content (MC). These parameters were specifically selected to investigate the effects of design (ST and CDS) and operating parameters (LAI, LR, SR and MC) on green roof thermal performance.

Simulations were conducted with EnergyPlus 8.3.0. for a period of 30 days in summer (December 2014). Each input parameter was varied for three to four values (Table 1). The EnergyPlus weather file for the simulations was prepared with the data collected by a weather station on top of the two storey Main Building at The University of Melbourne's Burnley campus (six kilometres from the centre of Melbourne). Data include ambient air temperature, ambient air relative humidity, rainfall, wind speed, wind direction and photosynthetically active radiation (PAR) collected every six minutes and averaged for one hour. Direct solar radiation was collected from another weather station located

500 m away from the roof weather station. Infrared downward radiation was calculated using the model developed by Bras [37]. The internal boundary conditions were set as 21 °C for heating and 20.9 °C for cooling.

Table 1. Input values for EnergyPlus simulations.

Input Variable/Parameter	Units	Input Values
Height of plants	m	0.20
Leaf area index	-	0.01–1–3–5
Leaf reflectivity	-	0.10–0.22–0.30–0.50
Leaf emissivity	-	0.95
Minimum stomatal resistance	s m^{-1}	50–150–180–30
Roughness	-	MediumRough
Thickness	m	0.07–0.10–0.15–0.30
Conductivity of dry soil	W m^{-1} K^{-1}	0.20–0.35–0.40–0.80
Density of dry soil	kg m^{-3}	1100
Specific heat of dry soil	J kg^{-1} K^{-1}	1200
Thermal infrared absorptance	-	0.90
Solar absorptance	-	0.70
Visible absorptance	-	0.75
Saturation volumetric moisture content of the soil	-	0.20–0.30–0.40
Residual volumetric moisture content of the soil	-	0.01
Initial volumetric moisture content of the soil	-	0.10
Moisture diffusion calculation method	-	Advanced

The energy fluxes (Equations (1)–(6)) in Sailor's model [28] were developed from FASST vegetation model by Frankenstein and Koenig [36]. Sailor's model highlights two main fluxes: one for the soil (substrate) layer (Equation (1)) and the other for the vegetation (Equation (4)). The sensible and the latent heat flux components incorporated in each main equation are explained in Equations (2), (3), (5) and (6).

Energy flux for the soil (W m^{-2}):

$$F_g = \left(1 - \sigma_f\right)\left[I_S^{\downarrow}\left(1 - \alpha_g\right) + \varepsilon_g I_{ir}^{\downarrow} - \varepsilon_g T_g^4\right] - \frac{\sigma_f \varepsilon_g \varepsilon_f \sigma}{\varepsilon_1}\left(T_g^4 - T_f^4\right) + H_g + L_g + K\frac{\delta T_g}{\delta z}, \tag{1}$$

Sensible heat flux at the foliage/soil interface (W m^{-2}):

$$H_g = \rho_{ag} C_{p,a} C_h^g W_{af}\left(T_{ag} - T_g\right), \tag{2}$$

Latent heat exchanges of the soil (W m^{-2}):

$$L_g = C_e^g l_g W_{af} \rho_{ag}\left(q_{af} - q_g\right), \tag{3}$$

Energy flux for vegetation (W m^{-2}):

$$F_f = \sigma_f\left[I_S^{\downarrow}\left(1 - \alpha_f\right) + \varepsilon_f I_{ir}^{\downarrow} - \varepsilon_f \sigma T_f^4\right] + \frac{\sigma_f \varepsilon_g \varepsilon_f \sigma}{\varepsilon_1}\left(T_g^4 - T_f^4\right) + H_f + L_f, \tag{4}$$

Sensible heat flux at the atmosphere/foliage interface (W m^{-2}):

$$H_f = 1.1 \cdot LAI\, \rho_{af} C_{p,a} C_f W_{af} \cdot \left(T_{af} - T_f\right), \tag{5}$$

Latent heat exchanges of the foliage (W m^{-2}):

$$L_f = l_f \cdot LAI \rho_{af} C_f W_{af} r''\left(q_{af} - q_{f,sat}\right), \tag{6}$$

2.2. Experimental Green Roofs

Computer simulations may provide inaccurate or, sometimes, unrealistic results when the conditions for the embedded assumptions are no longer valid. For this reason, we present substrate temperatures and heat fluxes collected from three experimental green roofs, and a bare conventional bituminous roof (no plants or substrates) on the Main Building at The University of Melbourne's Burnley Campus.

Each of the three green roofs has an approximate area of 15 m². They have a scoria mix (volcanic rock) substrate layer 100, 150 or 200 mm deep. Underneath the substrate layer, each green roof has:

- 0.6 mm filter layer (ZinCo filter sheet SF);
- 40 mm drainage layer (ZinCo Floradrain® FD 40-E);
- 5 mm protection layer (ZinCo SSM45 protection mat); and
- 0.36 mm high-density polyethylene (HDPE) root barrier.

The roof of the building is a 190 mm concrete slab lined with a waterproof bituminous coating and 10 mm of plaster board on the inside.

We first measured and recorded soil temperatures and heat fluxes from the green roofs in summer 2014 and winter 2015 before they were planted (non-vegetated), and then we collected data from the same green roofs when they were sparsely vegetated in summer 2015. Temperatures were measured with thermistors (Emerson Climate Technologies, Sidney, OH, USA, model 501-1125) placed at the surface and at the bottom of the substrate layer (Figure 1). Thermistors were placed in different locations across the three green roofs: four, five and six locations for the 100, 150 and 200 mm green roofs, respectively (Figure 2). Heat flux was measured in the centre of each roof with heat flux plate (Hukseflux, Delft, The Netherlands, model HFP01-L10m) placed between the green roof component layers and the top of the roof bituminous coating. The heat flux sensor for each roof, including the bare roof, was placed exactly at the middle to avoid edge influence. Data were recorded every six minutes by a Campbell Scientific data logger (CR1000-4M) and averaged over one hour.

After a data collection period with no vegetation (non-vegetated with substrate only), we planted each of the green roofs with three Australian high-water use plant species, *Stypandra glauca*, *Dianella admixta* and *Lomandra longifolia*. These plants were selected as part of a larger research project due to their high transpiration rates, but also because they can tolerate long drought periods common in Melbourne summers [38]. Plant foliage coverage was quantified through photo pixel counts using Adobe Photoshop CC 2015 program. Photos were taken by a GoPro Hero4 Camera (GoPro, Inc., San Mateo, CA, USA) 4 m above the central point of each green roof.

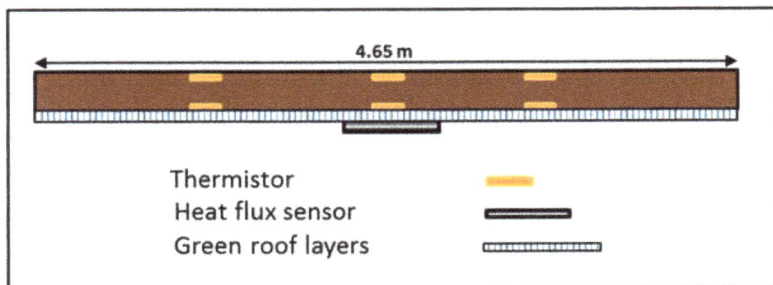

Figure 1. Hortizontal profile of the experimental green roofs at the University of Melbourne Burnley Campus illustrating sensor positions.

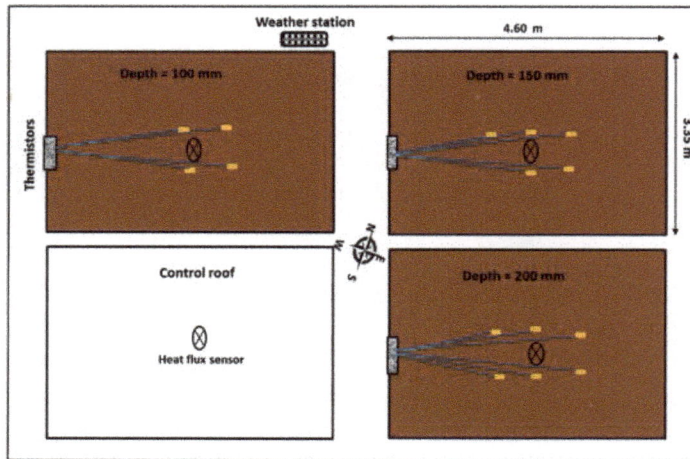

Figure 2. Design of the experimental green roofs at the University of Melbourne Burnley Campus illustrating the green and control roofs and sensor locations.

3. Results and Discussion

Results for the parametric analysis were simulated for 30 days in December 2014, but they are presented for only 10 days (2–11 December) in this paper. This was done primarily to aid interpretation through less condensed graphs. We selected two outputs from the simulations in EnergyPlus: the substrate temperature (°C) and the heat flux through the substrate (W m^{-2}) (soil, in the model). The outputs are calculated from Equations (1)–(3), which, by default, relate to the interface between the soil and the vegetation. The heat flux is calculated by adding the "green roof soil sensible heat transfer rate per unit area" to the "green roof latent heat transfer rate per unit area" outputs given by the simulations. We have selected these two outputs to compare the results with the measured data from the experimental green roofs. For the period simulated, there is an unrealistic result for the third day (Hours 49–72). As the substrate was initially very dry, justification for this result may be related to the rain event that occurred during Hours 46–55. The consequent rapid change in the substrate moisture content and the issues around the moisture diffusion accuracy [39], are likely to justify the unrealistic result simulated after the first rain event. However, this result does not hinder the main scope of this study. Future study will look further on model validation.

Substrate temperature and heat flux data were analysed from the three experimental green roofs during the same period (2–11 December 2014) for comparison.

Substrate temperature and heat flux data were analysed for 10 winter days (1–10 June 2015) to provide additional data from non-vegetated green roofs to better understand the effect of operating and design parameters during winter. Analyses for summer and winter also show rainfall, ambient air temperature and total incoming solar radiation.

Heat flux for 10 summer days (11–20 December 2015) when the roof was sparsely vegetated is also included. During this time, substrate temperatures were not collected. Heat flux of the bare roof was also collected and presented.

3.1. Parametric Analysis

Among the vegetation and operational parameters, the parametric analysis showed that significantly higher LAI values reduced both the substrate temperature (Figure 3) and the heat flux (Figure 4). This agrees with studies conducted in Mediterranean climate areas [20,21]. Reduction of the temperature was apparent during daytime and particularly for sunny days. The highest Leaf Area Index value (LAI = 5) reduced the temperature by up to 25 °C compared to the lowest value (LAI = 0.01) (i.e., Hours 1–24 and 169–192). On cloudy days (i.e., Hours 25–48 and 121–144), the magnitude of this parameter was considerably reduced. Even though LAI = 5 offered the greatest temperature reduction benefits, results for LAI = 3 were less than 5 °C higher than LAI = 5 at maximum. Results were similar on cloudy days. For heat flux (Figure 4), LAI = 3 and LAI = 5 offered a comparable result, meaning that a less dense vegetation can provide the same effect as very dense vegetation. However, LAI is normally slow to accumulate on hot and dry green roofs when plants are planted as tube stock, therefore additional irrigation and high density planting would be necessary to reach such high LAI values and maintain healthy and dense plants.

In contrast to LAI findings, the other two vegetation and operational parameters tested (SR and LR) and soil parameters did not show such a significant benefit in all the simulations conducted. For this reason, LR figures are not reported here, but they can be found in the conference paper [40] together with figures for ST and MC.

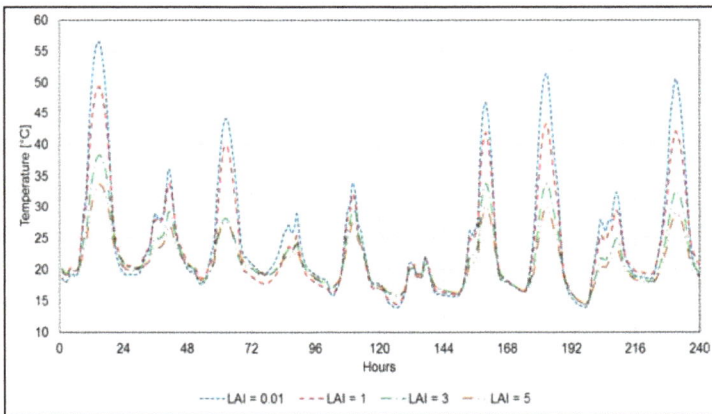

Figure 3. Parametric analysis using Sailor's (2008) green roof thermal model of soil temperature with varying Leaf Area Index (LAI) values for 10 summer days in Melbourne (2–11 December 2014).

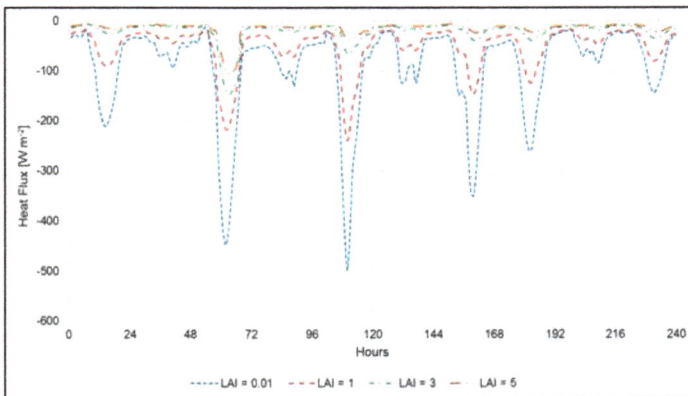

Figure 4. Parametric analysis using Sailor's (2008) green roof thermal model of soil heat flux with varying Leaf Area Index (LAI) values for 10 summer days in Melbourne (2–11 December 2014).

Simulations were done keeping LAI = 1 to better isolate the effect of Stomatal resistance (SR) (Figures 5 and 6) and Leaf reflectivity (LR). On some dry days (i.e., Hours 1–24 and 217–240, Figure 5), simulations of high SR values showed a considerable difference compared to simulations of low SR values. For example, substrate temperature of SR = 300 simulation was up to 4 °C higher than SR = 50 (Hours 217–240). This is because high stomatal resistance value translates into a low transpiration rates, as the plants attempt to conserve water, and thus have a lesser cooling effect. In addition, we can notice slight differences in Hours 50–80. They occur after a rainfall event followed by a hot day, with temperature close to 30 °C. Under this condition, plants with higher SR conserved the water in their leaves, and thus increased the temperature by significantly reducing transpiration. After this period, additional rain events occur, however the following days were not particularly hot and the moisture content is likely to be retained longer until Hours 200–240 when the plants reach a similar stress as explained before.

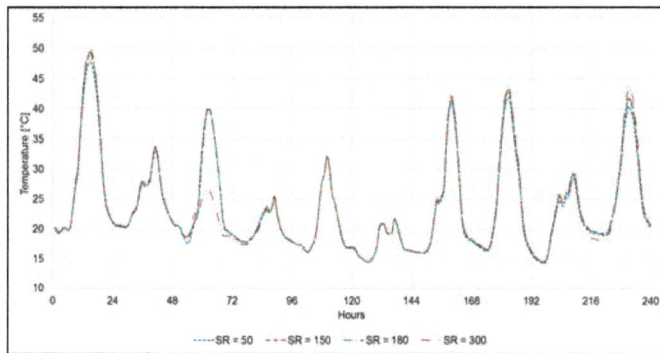

Figure 5. Parametric analysis using Sailor's (2008) green roof thermal model of soil temperature with varying minimum Stomatal Resistance (SR) [s m^{-1}] values for 10 summer days in Melbourne (2–11 December 2014).

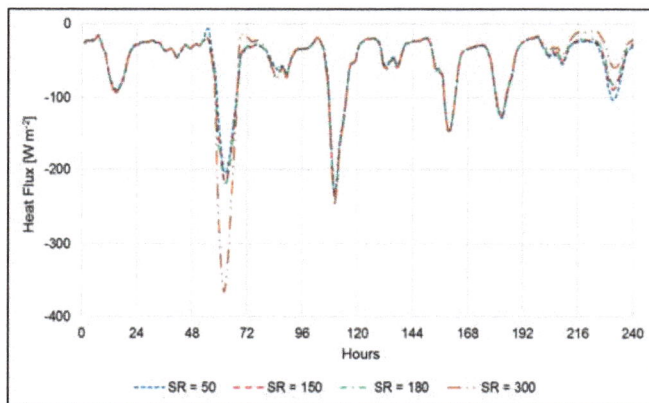

Figure 6. Parametric analysis using Sailor's (2008) green roof thermal model of soil temperature with varying minimum Stomatal Resistance (SR) [s m^{-1}] values for 10 summer days in Melbourne (2–11 December 2014).

None of the soil parameters considered significantly reduced soil temperature or the heat flux. However, this was expected because the simulations in EnergyPlus with Sailor's model only consider the interface between vegetation and soil; hence, they overlooked the benefits of the layers underneath this interface (i.e., the whole substrate layer, the drainage layer, etc.).

Surprisingly, there was little effect of varying the design parameter substrate thickness (ST) as simulation results were consistent for all values, except on two dry days (i.e., Hours 73–96 and 210–225 [40]) where the thickest simulated extensive green roof substrate (ST = 0.30 m) had cooler temperatures than the other values. Heat flux of the thickest substrate on these days was lower [40], meaning that larger thermal mass, when dry, can provide a better cooling effect.

The conductivity of dry soil (CDS) design parameter (Figures 7 and 8) showed varying results for most days across all simulations. CDS = 0.2 simulated temperatures up to 5 °C higher than CDS = 0.8 on sunny days (i.e., Hours 145–168 and 169–192), and provided similar temperatures as CDS = 0.35, CDS = 0.4 and CDS = 0.8 simulations on cloudy days (i.e., Hours 121–144, Figure 7). Similarly, heat flux (Figure 8) differences were more evident on sunny days, but did not have a clear pattern. Due to the limitations of the soil outputs, which we have already explained, the parametric analysis does not fully reveal the importance of the CDS for the green roof thermal performance as shown in previous studies [21,39,41].

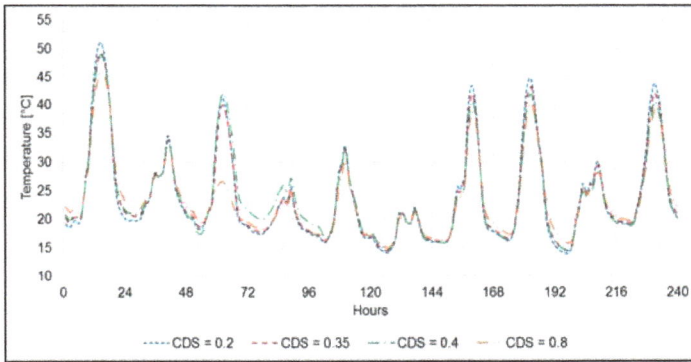

Figure 7. Parametric analysis using Sailor's (2008) green roof thermal model of soil temperature with varying Conductivity of Dry Soil (CDS) [W m^{-1} K^{-1}] values for 10 summer days in Melbourne (2–11 December 2014).

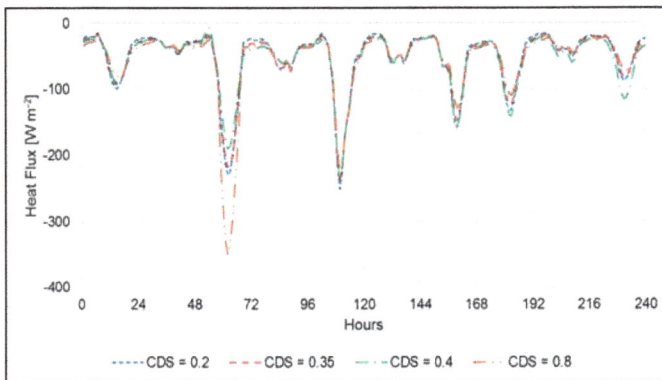

Figure 8. Parametric analysis using Sailor's (2008) green roof thermal model of soil heat flux with varying Conductivity of Dry Soil (CDS) [W m^{-1} K^{-1}] values for 10 summer days in Melbourne (2–11 December 2014).

Finally, the operational parameter soil moisture content (MC) showed small differences (~2 °C) in the temperature peaks on sunny days (i.e., Hours 169–192 [40]). The only significant differences were noted in the hours of a rainfall event (i.e., Hours 49 and 96 [40]). However, results were difficult to interpret for the soil temperature outputs, while they were generally uniform for the heat flux output [40].

Overall, the parametric analysis provided limited insight into the effects of the parameters tested, with the exception of the operational parameter LAI. To better understand the influence of different parameters, field measurements are recommended, in particular for those parameters that offered insignificant results, such as ST and MC. Examples are provided in the next section using results from experimental roofs in Melbourne.

3.2. Summer and Winter Temperatures and Heat Fluxes of Non-Vegetated Green Roofs

Temperatures and heat fluxes were measured on three non-irrigated and non-vegetated green roofs with depths of 100, 150 and 200 mm, and one bare conventional roof (heat flux only) in Melbourne over the same time period selected for the parametric analysis (2–11 December 2014). The period selected in this study represents typical days in Melbourne. They consist of rainy days followed by hot and sunny days within a relatively short period. This illustrates the various thermal characteristics of the experimental green roofs examined and their response to sudden changes. Additional temperatures and heat flux measurements are also provided for 10 days in the Australian winter, specifically from 1 to 10 June 2015. Temperatures collected at the bottom (B) and surface (S) of each green roof were hourly averaged for summer (Figure 9) and winter (Figure 10). Temperatures recorded at the surface of green roofs do not provide significant information on the benefits given by green roofs with different depths. Indeed, except for a limited number of small changes in the peak daily temperatures, the surface temperature trends of the three non-vegetated green roofs are generally identical. In contrast, heat fluxes and temperatures measured at the bottom of each non-vegetated green roofs have notable variation. For the purpose of this paper, we do not further analyse surface temperatures because compared to the temperatures at the bottom of a green roof profile, they do not provide significant evidence of the different green roof thermal performance.

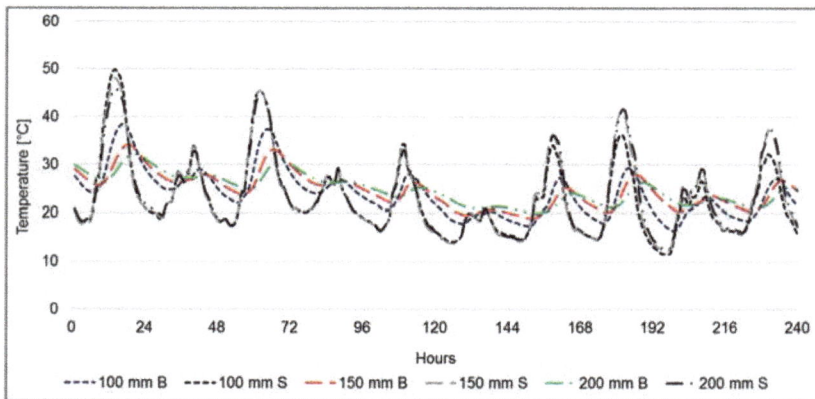

Figure 9. Green roof substrate temperature at the bottom (B) and at the surface (S) of three non-vegetated green roofs with different thickness for 10 summer days in Melbourne (2–11 December 2014).

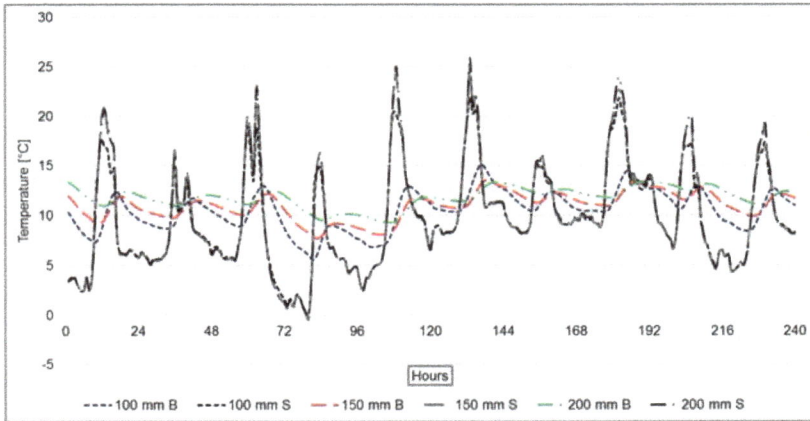

Figure 10. Green roof substrate temperature at the bottom (B) and at the surface (S) of three non-vegetated green roofs with different thickness for 10 winter days in Melbourne (1–10 June 2015).

3.2.1. Non-Vegetated Green Roof Thermal Performance in Summer

Figure 11 illustrates the temperature at the bottom of three non-vegetated green roofs 100, 150 and 200 mm thick. Temperature fluctuations were wider on sunny dry days (i.e., Hours 1–24 and 169–192) than cloudy rainy days (i.e., Hours 25–48 and 121–144) for every green roof. As expected, the 100 mm green roof had the largest fluctuations on all days ranging from 23.96 to 38.05 °C, compared to the 150 mm green roof, from 26.73 to 32.98 °C, and the 200 mm green roof, from 26.73 to 31.81 °C (Hours 1–24, Figure 11). The temperature difference between the 150 mm green roof and the 100 mm green roof was greater than between the 200 mm and 150 mm green roof, particularly on cloudy days. This suggested that the insulative effect of the additional thermal mass may increase until a green roof thickness of 150 mm and then may approach asymptote when the substrate is thicker than 150 mm.

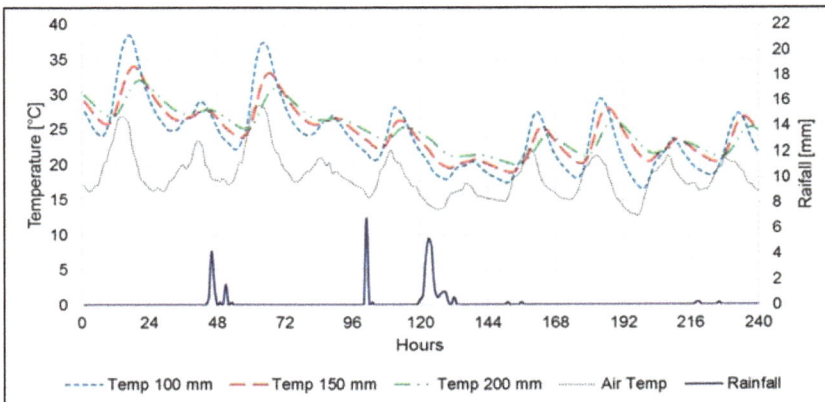

Figure 11. Rainfall and green roof substrate temperature at the bottom of three non-vegetated green roofs with different thickness for 10 summer days in Melbourne (2–11 December 2014).

Generally, the daily temperature peak was delayed in every green roof compared to the air temperature, and the thickest green roof had the lowest temperature peak, which was delayed up to four hours later than the thinnest green roof in this study. This was also confirmed by heat flux measurements from the green roofs at Burnley (Figures 12 and 13), where the 200 mm green roof delayed the heat flux peak up to eight hours compared to the heat flux of the bare roof (not shielded). As the heat flux measurements were collected at the interface between the building roof and the green roof, they took into account the effect of all the green roof component layers, and not only the substrate. Under this condition, the heat fluxes for the 150 mm and 200 mm green roofs were comparable, indicating that a 150 mm thick green roof may provide the same insulative effect as a thicker 200 mm deep green roof. The daily peak heat flux of the 100 mm green roof was double the other two green roofs (30 W m^{-2} vs. 15 W m^{-2}, Hours 1–24, 49–72 Figure 12). On cloudy and rainy days, however, the differences among the three green roofs were minimal. The thermal benefits from different substrate thicknesses and moisture contents were not reported on the temperatures measured at the surface layer or on the parametric analysis of the ST design parameter, as the soil temperature and heat flux were simulated by default at the interface between the substrate and the vegetation, rather than below all the green roof component layers. Simulations should be performed at the interface between the green roof and the roof building to take into account the thermal performance of the whole green roof profile.

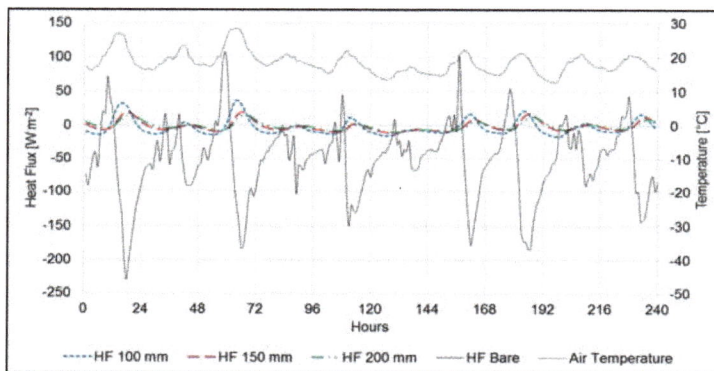

Figure 12. Ambient air temperature and heat flux of three non-vegetated green roofs with different thickness and one bare conventional roof for 10 summer days in Melbourne (2–11 December 2014).

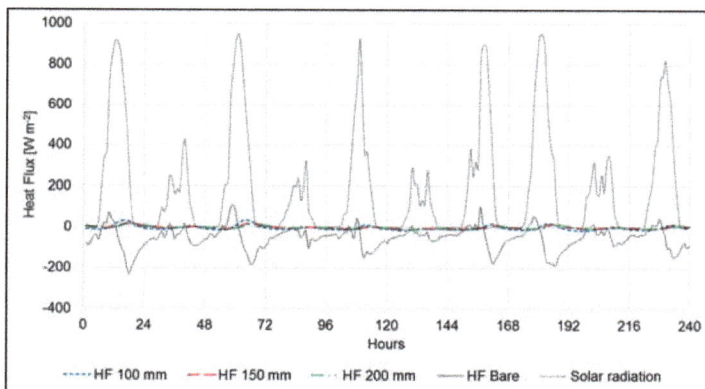

Figure 13. Solar radiation and heat flux of three non-vegetated green roofs with different thickness and one bare conventional roof for 10 summer days in Melbourne (2–11 December 2014).

3.2.2. Non-Vegetated Green Roof Thermal Performance in Winter

Winter data (Figures 14–16) had the same general trends as the summer measurement period, although with smaller temperature and heat flux differences among the three non-vegetated roofs. Rainy days did not smooth the temperatures, but instead increased the temperature variations: the 100 mm green roof recorded the lowest temperature (6.5 °C), while the 200 mm the highest (10.11 °C) on a rainy day (Hours 73–96, Figure 14). The heat flux values had comparable results for the 150 mm and 200 mm green roofs, although the heat flux for the 200 mm green roof was always negative on the tested days, and the one contributing the most at the peaks delay and reduction (Figures 15 and 16). The heat flux for the 100 mm green roof had always the greatest fluctuation, recording the lowest result in any day (i.e., -19.21 W m^{-2} Hour 82, Figure 15).

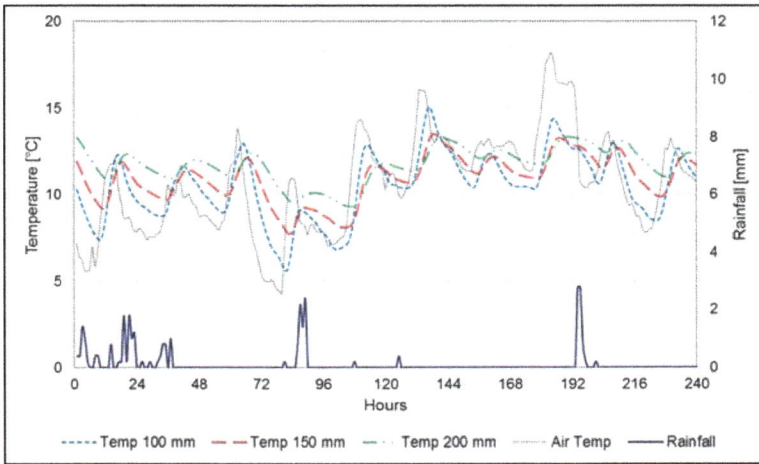

Figure 14. Rainfall and green roof substrate temperature at the bottom of three non-vegetated green roofs with different thickness for 10 winter days in Melbourne (1–10 June 2015).

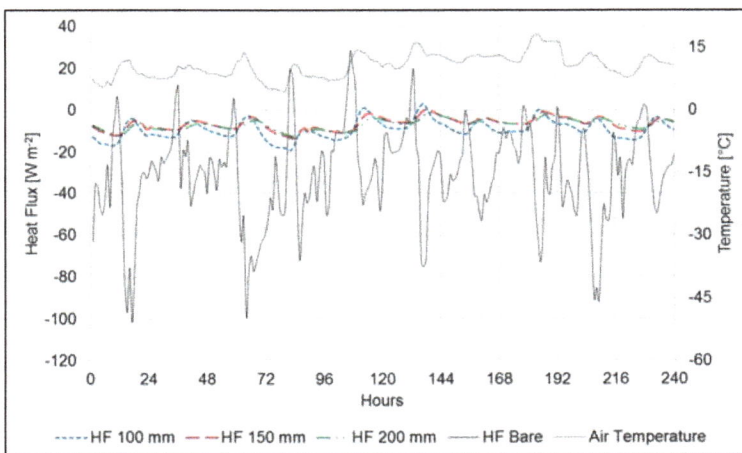

Figure 15. Ambient air temperature and heat flux of three non-vegetated green roofs with different thickness and one bare conventional roof for 10 winter days in Melbourne (1–10 June 2015).

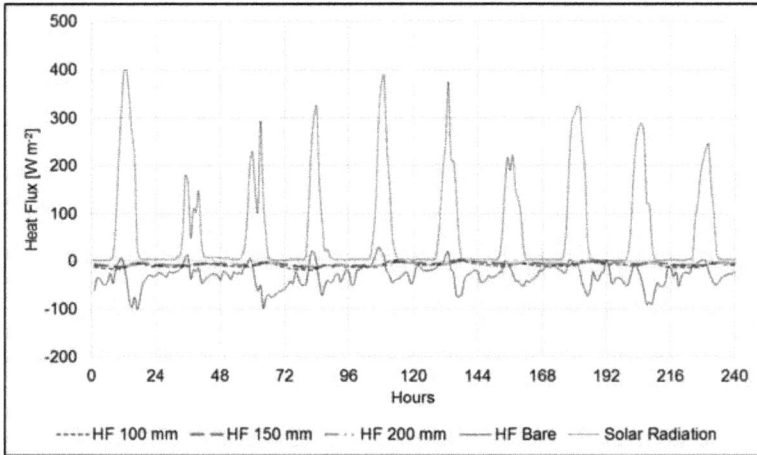

Figure 16. Solar radiation and heat flux of three non-vegetated green roofs with different thickness and one bare conventional roof for 10 winter days in Melbourne (1–10 June 2015).

In summary, we found that thicker substrates provided larger benefits for the thermal performance of green roofs by reducing and delaying up to eight hours the heat flux into the building to a greater extent. However, the delay effect of thicker substrates was not observed in the parametric analysis because the green roof substrate's temperature output overlooks the benefit of increasing thermal mass. The operational parameter MC, which changes after a rainfall event, influenced the substrate temperatures more in winter than in summer.

3.3. Summer Heat Fluxes of Sparsely Vegetated Green Roofs

The green roofs were planted at the end of October 2015, and the vegetation coverage in December 2015 was similar and around 20% for all the green roofs (Figures 17 and 18). As the plant species, density and size were the same for all of the plots, the associated LAI is also assumed similar. Establishment irrigation to field capacity (so that there was runoff) was provided three to five times a week for each green roof. Heat flux data were collected from 11 to 20 December 2015. Unfortunately, during summer 2015, temperatures for the green roof substrates were not collected. This period had mostly dry days, except for a few showers on the first (cumulative rain of 1.6 mm in a period of eight hours) and last day (cumulative rain of 4.4 mm in a period of 12 h); therefore, the green roofs were irrigated to retain high moisture content in the substrate, and maintain the plant health and growth. There was a heat wave event with maximum air temperatures between 35 °C and 41.5 °C for four consecutive days from 17 to 20 December.

Figure 17. The experimental green roofs at The University of Melbourne's Burnley Campus in December 2015 with sparse (20% cover) vegetation.

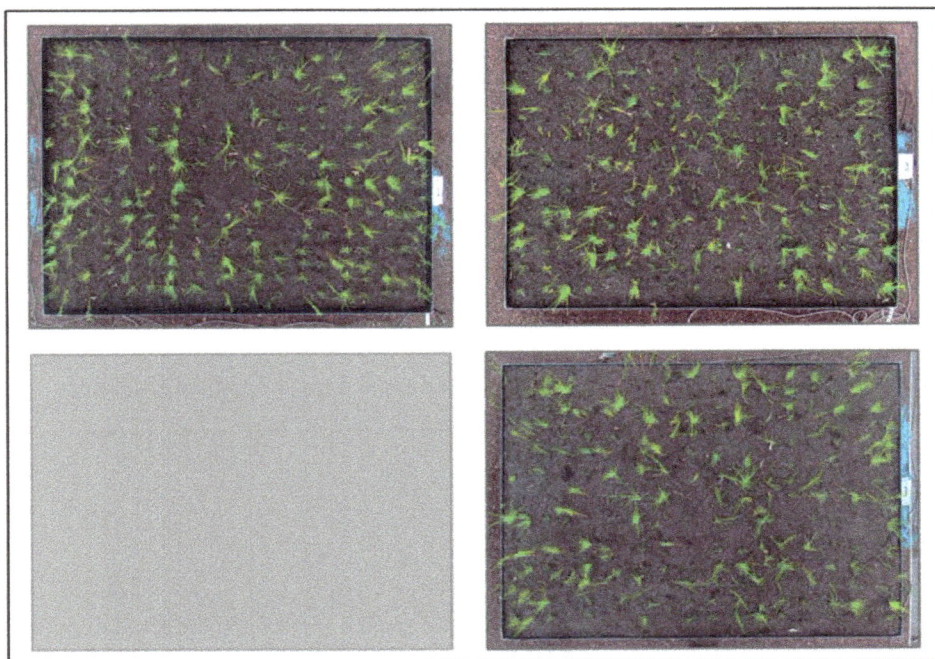

Figure 18. From top left corner clockwise, aerial view of vegetation coverage of 100, 150 and 200 mm sparsely vegetated experimental green roofs and reference concrete roof (Melbourne, December 2015).

In 2014, when the 200 mm and 150 green roofs were unvegetated and not irrigated, their heat fluxes were similar in magnitude and temporal pattern (Section 3.2.1). In 2015, when vegetated and irrigated, the peak heat flux of the 200 mm green roof (Figures 19 and 20) was distinctly lower than the 150 mm green roof (70% lower) and delayed about two hours (Hours 44–46, Figure 19). The difference between the two roofs is likely due to the large additional amount of irrigation water retained in the thicker 200 mm green roof and its larger cooling effect is likely due to both higher evaporation from the substrate and increased transpiration from the plants. In glasshouse conditions, Farrell et al. [38] found that these species increase water use according to water availability. However, due to the complexity of the evapotranspiration cooling effect of vegetated and irrigated green roofs in hot and dry climate, the authors understand that further study and research are necessary to better understand these dynamics.

Nonetheless, the heat flux difference between the 150 and 200 mm sparsely vegetated green roofs is minimal (i.e., heat peak flux 8.681 W m^{-2} versus 2.57 W m^{-2}, Hours 44–46, Figure 19) compared to the difference between the 100 and 150 mm sparsely vegetated green roof (i.e., heat peak flux 23.043 versus 8.681 W m^{-2}, Hours 42–44, Figure 19). This suggests again that the thermal performances of the 150 and 200 mm green roofs are overall comparable when all the green roof component layers (i.e., drainage layer, etc.) are included, as the heat flux are measured underneath the whole green roof. In addition, it suggests that the insulation thermal mass of the scoria substrate increases until a thickness of 150 mm, and then approaches the asymptote.

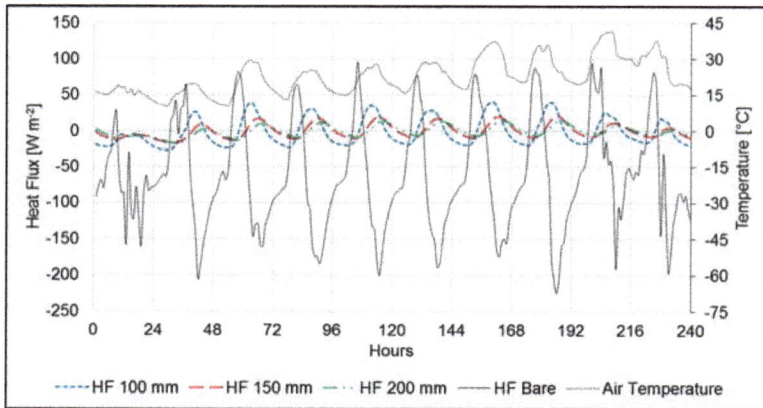

Figure 19. Ambient air temperature and heat flux of three sparsely vegetated green roofs with different thickness and one bare conventional roof for 10 summer days in Melbourne (11–20 December 2015).

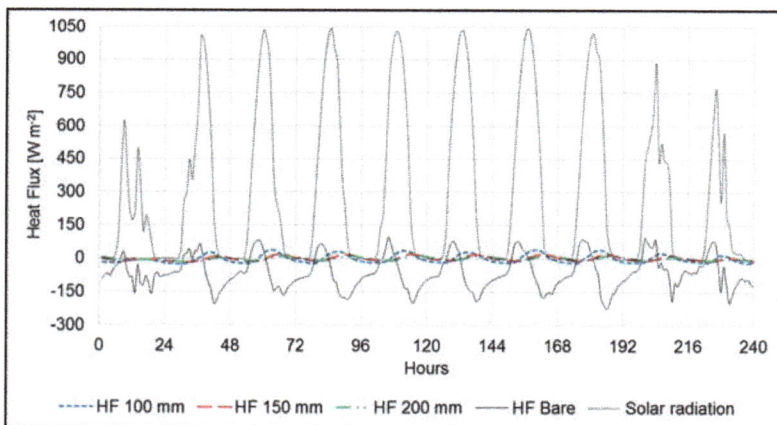

Figure 20. Solar radiation and heat flux of three sparsely vegetated green roofs with different thickness and one bare conventional roof for 10 summer days in Melbourne (11–20 December 2015).

Although the 200 mm green roof gives the best thermal performance, it has an analogous performance to the 150 mm green roof when unirrigated and unvegetated, and there is a small difference when it is irrigated and sparsely vegetated, but it is considerably heavier in weight. The 150 mm green roof with the scoria mixture substrate has a dead load of ~2 kN m^{-2} versus the 200 mm green roof which is ~2.5 kN m^{-2} [41,42]. Under this circumstance, a 50 mm shallower scoria substrate would have less weight (between 0.466 (dry) and 0.554 (saturated) kN m^{-2}) and would be a more viable option to retrofit to a wider range of existing buildings. While the extra dead load would be negligible for commercial buildings, this is not the case for most of Melbourne's low-rise residential buildings which have modest weight loading capacity [43]. If an alternative substrate were used, such as crushed roof tile [41,44], total load would increase up to ~3.15 kN m^{-2}, and it would be even more difficult to retrofit a residential building with a green roof [45].

4. Conclusions

Computer simulations using Sailor's green roof thermal model [28] were employed to investigate the effect of its design and operating parameters in Melbourne by means of parametric analysis. Furthermore, measured substrate temperatures and heat fluxes from non-vegetated and sparsely vegetated green roofs were collected to compare the effect of those parameters between experimental results and simulations.

In the parametric analysis, we found that LAI (for the vegetation layer) and, to a smaller extent, CDS (for the substrate layer) influenced green roof thermal performance the most. In particular for LAI, the parametric analysis suggests that LAI = 3 offers analogous results to a denser vegetation, such as LAI = 5. To reach such LAI values, irrigation and high density planting would be necessary to maintain plants health and dense cover. Computer simulations did not show much benefit from the other parameters we analysed, particularly those associated with the substrate.

In contrast, the measured data from the experimental roofs showed the importance of the substrate parameters. The green roof substrate alone considerably reduces the heat flux at the building-roof interface compared to a bare conventional roof and delays heat flux into the building by up to eight hours. The largest effect is found with the substrate thickness (ST), which shows significant temperature and heat flux reduction when varying from 100 to 150 mm. The computer simulations using Sailor's model do not reveal this benefit because the substrate's temperature and heat flux are simulated by default at the interface between substrate and vegetation, rather than to the interface between the building original roof and the green roof.

The thermal performance of green roofs with 150 and 200 mm thick substrates are generally comparable in summer and winter when non-vegetated and unirrigated and consistently higher than the green roof with 100 mm of substrate. When irrigated and sparsely vegetated, the 200 mm green roof provides the best thermal performance during summer. However, differences between the 200 mm and the 150 mm green roofs are minimal, while both have a notably greater thermal performance than the 100 mm green roof. Consequently, the 150 mm thick green roof would be a more widely applicable option for building retrofit due to its lighter weight than the 200 mm green roof (0.466–0.554 kN m^{-2} lighter).

As our study has further confirmed, the thermal performance of green roofs varies significantly across different climate zones and buildings. We have also highlighted how computer simulations may provide inaccurate or, sometimes, unrealistic results when the conditions for the embedded assumptions are no longer valid. As such, results from computer simulations should be validated with experimental data. This will help design and select green roof materials to maximise the green roof thermal performance in different climates and buildings.

Acknowledgments: This article is an extended version of the conference paper presented at ZEMCH 2016 International Conference in Kuala Lumpur, Malaysia, on 20–23 December 2016 [40]. This research was funded by Australian Research Council Linkage grant LP130100731 supported by Melbourne Water and Inner Melbourne Action Plan (IMAP) municipal councils. Andrea Pianella is supported by the Melbourne International Research Scholarship (MIRS), Melbourne International Fee Remission Scholarship (MIFRS) and The Frank Keenan Trust Fund Scholarship.

Author Contributions: Andrea Pianella conceived, designed and performed the experimental part and parametric analysis, analysed the data, structured the article and wrote the draft. Lu Aye suggested the parametric analysis. Lu Aye, Zhengdong Chen and Nicholas S. G. Williams provided comments to the experimental design, data analysis and amended the draft.

Conflicts of Interest: The authors declare no conflict of interest.

Nomenclature

$C_e{}^g$	bulk heat transfer coefficient for latent heat near the ground
C_f	bulk heat transfer coefficient for turbulent heat in the foliage
$C_h{}^g$	bulk heat transfer coefficient for sensible heat near the ground
$C_{p,a}$	specific heat of air at constant pressure
E_a	atmospheric emissivity
e	vapour pressure
$e_s(T_a)$	saturated vapour pressure at the air temperature
H_f	sensible heat flux at the atmosphere foliage interface [J m^{-2}]
H_g	sensible heat flux at the foliage/ground interface [J m^{-2}]
$I_{ir}{}^{\downarrow}$	total incoming infrared radiation [W m^{-2}]
$I_S{}^{\downarrow}$	total incoming solar radiation [W m^{-2}]
K	von Karmen constant
LAI	Leaf area index [m^2 m^{-2}]
L_f	latent heat exchanges of the foliage [J m^{-2}]
l_f	latent heat of vaporization at the foliage temperature [J kg^{-1}]
L_g	latent heat exchanges of the ground [J m^{-2}]
l_g	latent heat of vaporization at the ground temperature [J kg^{-1}]
q_{af}	mixing ratio of the air at the foliage interface
$q_{f,sat}$	saturated foliage mixing ratio
q_g	mixing ratio of the air at the ground surface
r''	$\frac{r_a}{r_a + r_s}$ surface wetness factor
r_a	aerodynamic resistance to transpiration [s m^{-1}]
r_s	foliage leaf stomatal resistance [s m^{-1}]
T_a	air temperature [K]
T_{af}	air temperature in the foliage [K]
T_f	temperature of foliage [K]
T_g	temperature of the ground surface [K]
W_{af}	wind speed in the foliage [m s^{-1}]
z	depth of the substrate [m]
α_f	shortwave albedo for the foliage [0–1]
α_g	shortwave albedo for the ground surface [0–1]
ε_1	$\varepsilon_f + \varepsilon_g - \varepsilon_f \varepsilon_g$
ε_f	longwave emissivity of the foliage [0–1]
ε_g	longwave emissivity of the ground surface [0–1]
ρ_{af}	density of air near the atmosphere/foliage interface [kg m^{-3}]
ρ_{ag}	density of air at the ground temperature [kg m^{-3}]
σ	Stefan-Boltzman constant

References

1. UNEP. The Emissions Gap Report. Are the Copenhagen Accord Pledges Sufficient to Limit Global Warming to 2° C or 1.5° C? A Preliminary Assessment. Available online: http://www.indiaenvironmentportal.org.in/files/The_EMISSIONS_GAP_REPORT.pdf (accessed on 16 August 2017).

2. Broto, V.C.; Bulkeley, H. A survey of urban climate change experiments in 100 cities. *Glob. Environ. Chang.* **2013**, *23*, 92–102. [CrossRef] [PubMed]

3. Doulos, L.; Santamouris, M.; Livada, I. Passive cooling of outdoor urban spaces. The role of materials. *Sol. Energy* **2004**, *77*, 231–249.

4. Omer, A.M. Energy, environment and sustainable development. *Renew. Sustain. Energy Rev.* **2008**, *12*, 2265–2300. [CrossRef]

5. Sadineni, S.B.; Madala, S.; Boehm, R.F. Passive building energy savings: A review of building envelope components. *Renew. Sustain. Energy Rev.* **2011**, *15*, 3617–3631. [CrossRef]

6. Chwieduk, D. Towards sustainable-energy buildings. *Appl. Energy* **2003**, *76*, 211–217. [CrossRef]

7. John, G.; Clements-Croome, D.; Jeronimidis, G. Sustainable building solutions: A review of lessons from the natural world. *Build. Environ.* **2005**, *40*, 319–328. [CrossRef]

8. Baik, J.J.; Kwak, K.-H.; Park, S.-B.; Ryu, Y.-H. Effects of building roof greening on air quality in street canyons. *Atmos. Environ.* **2012**, *61*, 48–55. [CrossRef]

9. Lee, K.E.; Williams, K.J.H.; Sargent, L.D.; Farrell, C.; Williams, N.S.G. Living roof preference is influenced by plant characteristics and diversity. *Landsc. Urban Plan.* **2014**, *122*, 152–159. [CrossRef]

10. Stovin, V. The potential of green roofs to manage urban stormwater. *Water Environ. J.* **2010**, *24*, 192–199. [CrossRef]

11. Gill, S.E.; Handley, J.F.; Ennos, A.R.; Pauleit, S. Adapting cities for climate change: The role of the green infrastructure. *Built Environ.* **2007**, *33*, 115–133. [CrossRef]

12. Williams, N.S.G.; Lundholm, J.; Scott Macivor, J. Do green roofs help urban biodiversity conservation? *J. Appl. Ecol.* **2014**, *51*, 1643–1649. [CrossRef]

13. Berardi, U.; Ghaffarian Hoseini, A.; Ghaffarian Hoseini, A. State-of-the-art analysis of the environmental benefits of green roofs. *Appl. Energy* **2014**, *115*, 411–428. [CrossRef]

14. Oberndorfer, E.; Lundholm, J.; Bass, B.; Coffman, R.R.; Doshi, H.; Dunnett, N.; Gaffin, S.; Köhler, M.; Liu, K.K.Y.; Rowe, B. Green roofs as urban ecosystems: Ecological structures, functions, and services. *BioScience* **2007**, *57*, 823–833. [CrossRef]

15. Wang, Y.; Berardi, U.; Akbari, H. Comparing the effects of urban heat island mitigation strategies for Toronto, Canada. *Energy Build.* **2016**, *114*, 2–19. [CrossRef]

16. La Roche, P.; Berardi, U. Comfort and energy savings with active green roofs. *Energy Build.* **2014**, *82*, 492–504. [CrossRef]

17. Ramesh, T.; Prakash, R.; Shukla, K.K. Life cycle energy analysis of buildings: An overview. *Energy Build.* **2010**, *42*, 1592–1600. [CrossRef]

18. La Roche, P.; Carbonnier, E.; Halstead, C. Smart green roofs: Cooling with variable insulation. In Procceedings of the PLEA2012-28th Conference, Opportunities, Limits & Needs towards an Environmentally Responsible Architecture, Lima, Peru, 7–9 November 2012.

19. Pianella, A.; Bush, J.; Chen, Z.; Williams, N.S.G.; Aye, L. Green roofs in Australia: Review of thermal performance and associated policy development. In Proceedings of the Architectural Science Association Conference, Adelaide, Australia, 7–9 December 2016; Zuo, J., Daniel, L., Soebarto, V., Eds.; pp. 795–804.

20. Bevilacqua, P.; Mazzeo, D.; Bruno, R.; Arcuri, N. Experimental investigation of the thermal performances of an extensive green roof in the Mediterranean area. *Energy Build.* **2016**, *122*, 63–79. [CrossRef]

21. Olivieri, F.; Di Perna, C.; D'Orazio, M.; Olivieri, L.; Neila, J. Experimental measurements and numerical model for the summer performance assessment of extensive green roofs in a Mediterranean coastal climate. *Energy Build.* **2013**, *63*, 1–14. [CrossRef]

22. Fioretti, R.; Palla, A.; Lanza, L.G.; Principi, P. Green roof energy and water related performance in the Mediterranean climate. *Build. Environ.* **2010**, *45*, 1890–1904. [CrossRef]

23. Bevilacqua, P.; Mazzeo, D.; Bruno, R.; Arcuri, N. Surface temperature analysis of an extensive green roof for the mitigation of urban heat island in southern mediterranean climate. *Energy Build.* **2017**, *150*, 318–327. [CrossRef]

24. Liu, K.; Baskaran, B. Thermal performance of green roofs through field evaluation. In Proceedings of the First North American Green Roof Infrastructure Conference, Awards and Trade Showm, Chicago, IL, USA, 29–30 May 2003.

25. Jim, C.Y.; Tsang, S.W. Biophysical properties and thermal performance of an intensive green roof. *Build. Environ.* **2011**, *46*, 1263–1274. [CrossRef]

26. Jim, C.Y. Effect of vegetation biomass structure on thermal performance of tropical green roof. *Landsc. Ecol. Eng.* **2012**, *8*, 173–187. [CrossRef]

27. Schweitzer, O.; Erell, E. Evaluation of the energy performance and irrigation requirements of extensive green roofs in a water-scarce Mediterranean climate. *Energy Build.* **2014**, *68*, 25–32. [CrossRef]

28. Sailor, D.J. A green roof model for building energy simulation programs. *Energy Build.* **2008**, *40*, 1466–1478. [CrossRef]
29. Brown, C.; Lundholm, J. Microclimate and substrate depth influence green roof plant community dynamics. *Landsc. Urban Plan.* **2015**, *143*, 134–142. [CrossRef]
30. Dunnett, N.; Nagase, A.; Hallam, A. The dynamics of planted and colonising species on a green roof over six growing seasons 2001–2006: Influence of substrate depth. *Urban Ecosyst.* **2008**, *11*, 373–384. [CrossRef]
31. Lundholm, J.T. Spontaneous dynamics and wild design in green roofs. *Isr. J. Ecol. Evol.* **2016**, *62*, 23–31. [CrossRef]
32. Palomo Del Barrio, E. Analysis of the green roofs cooling potential in buildings. *Energy Build.* **1998**, *27*, 179–193. [CrossRef]
33. Tabares-Velasco, P.C.; Srebric, J. A heat transfer model for assessment of plant based roofing systems in summer conditions. *Build. Environ.* **2012**, *49*, 310–323. [CrossRef]
34. Lazzarin, R.M.; Castellotti, F.; Busato, F. Experimental measurements and numerical modelling of a green roof. *Energy Build.* **2005**, *37*, 1260–1267. [CrossRef]
35. Ouldboukhitine, S.E.; Belarbi, R.; Jaffal, I.; Trabelsi, A. Assessment of green roof thermal behavior: A coupled heat and mass transfer model. *Build. Environ.* **2011**, *46*, 2624–2631. [CrossRef]
36. Frankestein, S.; Koeing, G. *FASST Vegetation Models, Technical Report TR-04-25*; U.S. Army Engineer Research and Development Center: Vicksburg, MS, USA, 2004.
37. Bras, R.L. *Hydrology: An Introduction to Hydrologic Science*; Addison-Wesley: Reading, MA, USA, 1990.
38. Farrell, C.; Szota, C.; Williams, N.S.G.; Arndt, S.K. High water users can be drought tolerant: Using physiological traits for green roof plant selection. *Plant Soil* **2013**, *372*, 177–193. [CrossRef]
39. Vera, S.; Pinto, C.; Tabares-Velasco, P.C.; Bustamante, W.; Victorero, F.; Gironás, J.; Bonilla, C.A. Influence of vegetation, substrate, and thermal insulation of an extensive vegetated roof on the thermal performance of retail stores in semiarid and marine climates. *Energy Build.* **2017**, *146*, 312–321. [CrossRef]
40. Pianella, A.; Aye, L.; Chen, Z.; Williams, N.S.G. Effects of design and operating parameters on green roof thermal performance in Melbourne. In *Proccedings of the 5th International Conference on Zero Energy Mass Customised Housing (ZEMCH)*, Kuala Lumpur, Malaysia, 20–23 December 2016; Hashemi, A., Ed.; pp. 11–31.
41. Pianella, A.; Clarke, R.E.; Williams, N.S.G.; Chen, Z.; Aye, L. Steady-state and transient thermal measurements of green roof substrates. *Energy Build.* **2016**, *131*, 123–131. [CrossRef]
42. Victorian Department of Environment and Primary Industries. *Growing Green Guide: A Guide to Green Roofs, Walls and Facades in Melbourne and Victoria, Australia*; Victorian Department of Environment and Primary Industries: Melbourne, Australia, 2014.
43. Sofi, M.; Zhong, A.; Lumantarna, E.; Cameron, R. Addition of green: Re-evaluation of building structural elements. In *Proceedings of the Practical Responses to Climate Change Conference*, Melbourne, Australia, 25–27 November 2014; Engineers Australia: Barton ACT, Australia; pp. 100–107.
44. Farrell, C.; Mitchell, R.E.; Szota, C.; Rayner, J.P.; Williams, N.S.G. Green roofs for hot and dry climates: Interacting effects of plant water use, succulence and substrate. *Ecol. Eng.* **2012**, *49*, 270–276. [CrossRef]
45. Wilkinson, S.; Feitosa, R.C.; Kaga, I.T.; deFranceschi, I.H. Evaluating the Thermal Performance of Retrofitted Lightweight Green Roofs and Walls in Sydney and Rio de Janeiro. *Procedia Eng.* **2017**, *180*, 231–240. [CrossRef]

sustainability

MDPI

Article

Adoption of Energy Design Strategies for Retrofitting Mass Housing Estates in Northern Cyprus

Bertug Ozarisoy [1],* and Hasim Altan [2]

[1]   School of Architecture, Computing & Engineering, University of East London, London E16 2RD, UK
[2]   Department of Architectural Engineering, University of Sharjah, Sharjah 27272, UAE; hasimaltan@gmail.com
*   Correspondence: ozarisoyb@gmail.com; Tel.: +44-7-961-974-826

Received: 28 June 2017; Accepted: 13 August 2017; Published: 21 August 2017

Abstract: This research project is undertaken in the Turkish Republic of Northern Cyprus (T.R.N.C.). The objective of the research is to investigate the occupants' behaviour and role in the refurbishment activity by exploring how and why occupants decide to change building systems and how to understand why and how occupants consider using energy-efficient measurements. The housing estates are chosen from 16 different projects in four different regions of the T.R.N.C. that include urban and suburban areas. The study is conducted through semi-structured interviews to identify occupants' behaviour as it is associated with refurbishment activity. This paper presents the results of semi-structured interviews with 70 homeowners in a selected group of 16 housing estates in four different parts of the T.R.N.C. Alongside the construction process and its impact on the environment, the results point out the need for control mechanisms in the housing sector to promote and support the adoption of retrofit strategies and to minimise non-controlled refurbishment activities. The results demonstrate that European Union Energy Efficiency directives need not only inform households about technological improvements that can be installed in their residential properties, but should also strongly encourage and incentivise them to use them efficiently. Furthermore, the occupants' energy consumption behaviour and the applicable policy interventions will make the difference between implementing policy which in fact delivers on its aims for energy efficiency and sustainability.

Keywords: construction process; energy efficiency; refurbishment activity; retrofitting; Cyprus

1. Introduction

This research project is undertaken in the Turkish Republic of Northern Cyprus (T.R.N.C.). This research investigates the socio-political developments that have had an impact on the architecture and the urban planning process in this particular region. The study focuses on identifying refurbishment activities capable of diagnosing and detecting the underlying problems alongside the challenges offered by the mass housing estates design and planning in addition to identifying the cultural influences in the refurbishment process, which allow for the maximisation of expected energy savings. The rapid construction activities are responsible for the consumption of approximately two-thirds of global energy demand in urban and suburban areas, and are therefore responsible for major changes in the built environment [1]. The Intergovernmental Panel on Climate Change report in 2007 indicates that urbanisation has led to an increase in temperatures of $0.006\ °C$ per decade since 1900 on the global land record and $0.002\ °C$ on the global and ocean record [2]. In the T.R.N.C. the increasing number of construction activities has had an impact on the environment, which is included in future assessments of problems for the mass housing sector. At the same time, as a result of an increase in summer temperatures and a decrease in temperatures during winter months have brought changes in urban energy use. The Ministry of Environment and Natural Resources Department of Meteorology—T.R.N.C. (Cevre ve Dogal Kaynaklar Bakanligi Meteoroloji Dairesi Mudurlugu in Turkish) statistics in 2015 shows that the average annual temperature was 17.2 °C between 1960 and 1991 and this increased to 17.7 °C between 1991 and 2007 [3]. These results show that the island is threatened by the climate change impact now affecting the whole planet but within the T.R.N.C.

The rapid construction during the "property boom" years led to a revived interest in the property market. The expectations of the Annan Plan and changing market conditions throughout the world is prominent evidence that people from countries such as Russia, Turkey, Greece, the United Kingdom, and Germany began to show significant interest in buying their "second homes" in the T.R.N.C. [4]. The increasing energy demand by the residential sector was felt mainly through rapid construction activities and a renewed concentration on economic improvement. In the T.R.N.C., the rapid and varied construction activity throughout the building sector resulted in economic growth. The State Planning Organisation—T.R.N.C. (Devlet Planlama Orgutu—K.K.T.C. in Turkish) in 2008 statistics show that, in the pre-construction period between 1997 and 2001, the GNP rate had an average of 1.8% [5]. However, during the accelerated construction activity period between 2002 and 2006, this rate had jumped to 11% per annum. It should also be noted that during this same time the construction industry accounted for 8.1% of GNP in the T.R.N.C. [5]. The results show how construction activity activated interest in construction projects. However, the situation led to unsustainable environmental problems, ecological constraints and energy issues.

The solutions of retrofit strategies should be effective, environmentally acceptable and feasible given the type of mass housing projects under review, with due regard for their location, the climatic conditions, within which they are undertaken, the socio economic standing of the house owners and their cultural assessments, local resources and legislative constraints. Furthermore, the study goes on to insist on the practical and long-term economic benefits of implementation of retrofit strategies under the selected research methodology (ethnographic study) and why this should be fully understood by the construction companies and householders.

The literature review has been obtained by a combination of descriptive and explanatory research methods to inform the research background and elaborate the justification of the research context on the researcher's knowledge and experiences. For this purpose, the research carries out a literature review on these key-drivers to support the theoretical framework and give information on the construction industry and its process in the T.R.N.C. It details the energy consumption of the residential buildings under review, as well as the energy efficiency and retrofit strategies in order to understand how the European Union (EU) objectives are regulating the housing sector as a way to improve conditions both in policies and practices in the residential sector. In these theoretical foundation methods, the literature survey has been carried out through a collection of periodicals (i.e., reports, journals, articles, books) on the European Union energy efficiency standards, implementations, policy documents and the related sources on the retrofit practice by other EU member states have been studied to understand the role of EU objectives, allowing for a better understanding for the identification of the research problems. The subsequent sections in this paper are structured as follows; the paper will first discuss the background and justification of research, followed by the hypothesised relationships with regard to the relevant literature. This is then continued with explanations on the methodology employed. Preliminary findings and discussions are given prior to the conclusion. Limitations and future research directions will also be discussed.

2. Location and Climate

Cyprus, located in the eastern Mediterranean, is the third largest island after Sicily and Sardinia. It is located in the Eastern Mediterranean part. Its closest neighbours are Turkey to the north and Lebanon, Syria, Israel, Egypt and Greece to the south and southwest. It sits on latitude 35° N and longitude 33° E (Table 1). It is the most eastern member of the European Union after being admitted as an EU member on 1 May 2004. According to the Cyprus Meteorological Service data, in 2013, the main geomorphological characteristics of the island are as follows: coastal climatic zone, inland climatic zone, semi-mountainous climatic zone and mountainous climatic zone (Figure 1) [6].

Figure 1. The geographical map of Cyprus.

The Cyprus Meteorological Service data in 2013 indicates that, generally, in July and August the mean daily temperature rises by approximately 30 °C on the central plain [6]. According to the same report, the mean daily temperature is recorded at 23 °C on the Troodos mountains. It is also noted that the average maximum temperatures are 37 °C and 28 °C, respectively. It is important to say that these temperature rates are not so bad for summertime. However, in July and August, there are a few days where temperatures can reach up to 45 °C inland and 40 °C on the coast [6]. Temperatures in the mountainous regions are much cooler than other regions during summertime (Table 1).

Temperatures in winter in Cyprus are 3–10 °C on the central plain and 0–5 °C on the higher parts of the Troodos Mountains. Naturally, temperatures on lower ground and the coastline are significantly lower, ranging between 8 °C and 12.5 °C [6]. In general, and using the available data, Cyprus has relatively mild winter conditions. However, there are occasions during the winter period when temperatures may drop below 0 °C, especially at higher altitudes in the Troodos mountains and sometimes inland. Very low temperatures are unusual near the coast where they tend to be further down the temperature scale, even 2–3 °C on the coldest of winter nights.

Table 1. Geographical data and boundary conditions assumed.

Geographical Data			
Place	Altitude	Latitude	Longitude
Cyprus	11952 m	35° N	33° E
Thermal Data			
Degree day: 2259 Average heating period temperature: 7.3 °C		Climatic Area: Average solar radiation on a horizontal plane during the heating period: 7.92 MJ/m^2	
Summer Condition Data			
Month of max solar radiation: July Summer max temperature: 45 °C		Average monthly summer max temperature: Difference in temperature during the hottest day: 12 °C	

3. Construction Industry and its Impacts on Energy Use

In developing countries where urban growth and rapid urbanisation are occurring, uncontrolled urban sprawl, poor land use planning and poorly built housing estates has led to an impact on the current state of urbanisation and growth [7]. Hence, regarding the T.R.N.C. case, changing the physical layout of the land together with un-planned land use are two major factors, which have resulted in architectural, urban and environmental devastation (Figure 2). For instance, construction companies started their invasive developments in many cases without any official permission in the virgin shorelines, mountain regions and riverbeds and also before laying down any ground infrastructures such as roads, water, and electricity. This situation has prevented efficient services being made available to the project sites for their completion; therefore, it has resulted in the abandonment of the mass housing estates by the construction companies.

(a)   (b)

Figure 2. Invasive (**a**); and destructive (**b**) mass housing development of the untouched natural habitat.

In the T.R.N.C., urbanisation started in the 1980s because of the development in the economy, which prompted a simultaneous demand in the mass housing sector. This led to rapid construction of the apartment blocks, detached, semi-detached and terraced houses being built randomly across the country in both urban and suburban areas. As a result of this exponential growth in the property market, there is no political agenda for controlling urban planning, infrastructure and the physical quality of the building and its adaptability to the local environmental climate. This led to poorly built houses without any initiative in the reduction of energy consumption from the buildings.

One of the main principal problems in evaluating the energy performance of the recently built housing stock is represented by the lack of current building regulations in the Town Planning Law 55/89 (Sehir Planlama Yasasi in Turkish). The current policies are adopted from similar regulations left over from the British administration [8]. The Town Planning Law no longer reflects the need and priorities of today's development of urban and suburban areas. Because of the structure of the Town Planning Law 55/89, the problems of its existing poor urbanism approach in planning concerns are now an on-going hindrance to the introduction and enforcement of proper architectural design tools and control mechanisms in the construction of buildings [9].

The following research, which has been published previously, indicates that there is a lack of awareness in understanding the importance of energy use. One strategy for reducing this deficiency in understanding is to explain the variance in energy performance in terms of the gap between design and construction process (Figure 3) [10]. Furthermore, the identification of the building diagnosis varies according to the age, size, type, etc. of building [11].

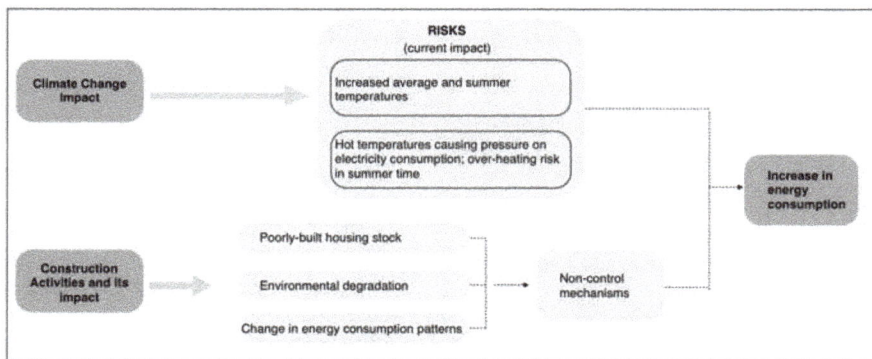

Figure 3. Model definition for justification of the research gap.

Recent studies indicate that the evolution of mass housing estates after the property boom in the T.R.N.C. brought massive changes in construction strategy, which now responds less favourably to occupants' requirements and also their current social and economic aspects [12–15]. To reverse the above man-made problems, an improvement in the physical quality of building stock is directly related to demands such as a reduction in energy consumption and thus reduction in carbon dioxide emissions [16]. In this study, the one main point is for the construction companies to assess and adopt the necessary principles of retrofit strategies to the present mass housing stock to bring into effect the above stated matters.

The approach here is to look at buildings that have been built by privately owned construction companies and have already been retrofitted by occupants to make the building more energy efficient and adaptable to the local environment. This research is prompted by a recognition that the current planning policies have not been effective in taking into account the energy consumption of the recently built mass housing estates by the construction companies in the T.R.N.C. between 2003 and 2015 (the property boom occurred during this particular period because of the political changes in Cyprus). This research reveals that there is an urgent need for the governmental bodies to bring out new and effective polices for the mass housing sector to force the construction industry to apply the necessary retrofit strategies on a rapid and large-scale basis to reduce energy consumption.

4. Current Energy Efficiency Awareness

To put the question of energy efficiency in buildings in the T.R.N.C. into context, former and current energy policies, as well as the current energy consumption situation is presented in Figure 4 (Phase I & II).

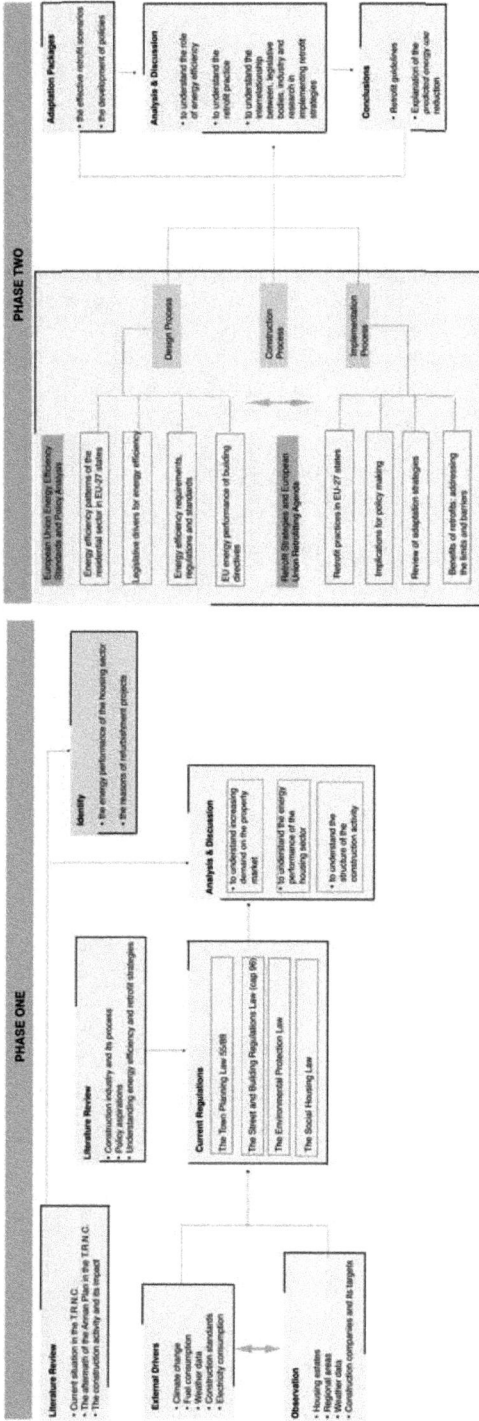

Figure 4. The collected and reviewed data on current regulations in the T.R.N.C.

4.1. Energy Consumption and Policies

The research concept sits within the context of concerns both over climate change owing to anthropogenic emissions of carbon dioxide and associated greenhouse gases and over future energy security owing to depletion of fossil fuel reserves. In accordance to the statement, high energy consumption creates serious problems in the European Union member states. Buildings are responsible for 40% of world energy consumption [17]. The Electricity Authority of the T.R.N.C. (KIB-TEK—Kibris Turk Elektrik Kurumu in Turkish) in 2017 indicates that residential sector consumption consumed 230.367 MkWh (Million kilowatt-hours) in 2003 and this figure rose to 377.971 MkWh in 2015 (Figure 5) [18].

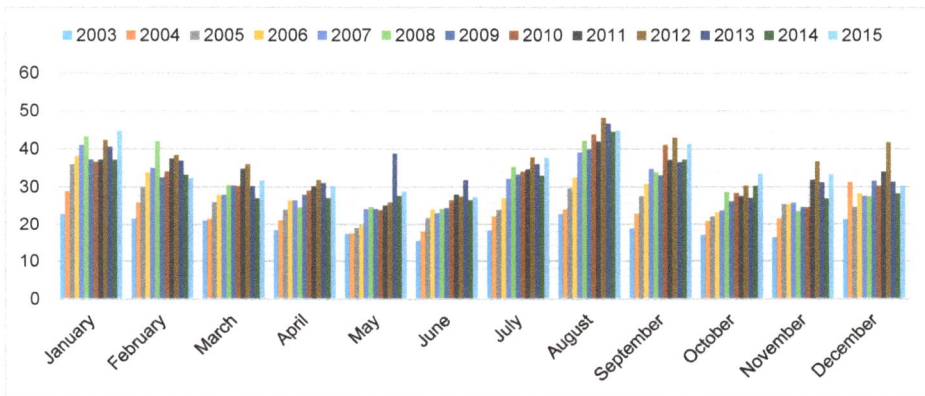

Figure 5. The energy consumption of residential buildings between 2003 and 2015.

The need for convergence between the predicted and the actual energy consumption, in a high energy performance building, is an essential factor in the design and construction process. For this reason, the recast of the direction on the energy performance of buildings 2010/31 requires the adoption of an action plan for the significant decrease of the energy consumption of buildings by 2020 [19]. However, the T.R.N.C is an isolated country internationally, for political and historical reasons, and its exclusion from European directives has meant that the action plan has not been implemented by governmental bodies.

In a similar way, the Republic of Cyprus has already implemented energy efficiency requirements and issued energy certificates for their own construction industry under the European Union objectives. It has also started paying more attention to the monitoring of the housing stock and promoting energy saving strategies to the occupants. This process is intended to match the standards demanded by other European states and is, at the same time, a recognition of the absolute need to improve their housing conditions. Apart from the issues concerning the energy use of buildings within the implementation bodies, are the European Union objectives being introduced to highlight the significance of control mechanisms in the area of the housing sector and the importance of prioritising areas on the outskirts of urban areas and suburban areas in transition and areas of environmental importance.

Some current and previous research has been conducted on energy efficiency and other forms of energy saving in housing. A considerable amount of literature has been published on energy-saving measures and strategies and their adaptability to individual dwellings (e.g., solar panels, types of insulation, and high-efficiency boilers), apartment blocks (e.g., heating and cooling demand) or neighbourhoods (e.g., district heating). What we know about energy consumption reduction is largely based upon empirical studies that investigate how government or institutions pay attention to implementing energy-efficiency regulations, including taxes and subsidies. Additionally, several studies have documented that there is a lack of control mechanisms between energy regulations and implementation process [20–26]. Recently, researchers have shown an increased interest in assessing tools for sustainability or, conversely, the environmental impact of buildings such as employing "life cycle assessment" principles [27–29]. However, the issue of embedding of energy efficiency has received considerable critical attention in the management of private institutions, organisations or privately owned construction companies.

The improvement of quality of the building components of the existing housing stock has been a matter of concern in housing energy management and improving energy efficiency in the T.R.N.C. Although the energy performance of existing building stock has not been recognised as a research subject in this context, the importance of implementing energy efficiency regulations and installation of energy-efficient technologies has considerably increased in the last decade in EU member states. It is also worth mentioning that energy efficiency in the built environment is now a frequently debated subject, particularly in Europe and in the UK, explored in many scholarly and professional publications. However, a lack of research has been conducted in terms of understanding energy performance of existing buildings and thermal comfort level conditions of occupants' in the residential buildings in the T.R.N.C.

4.2. Policy Initiatives for Reducing Energy Use in the Residential Sector

For EU and its member states, energy has become a significant issue. It has assumed a priority status in terms of its urgency as a problem and the action plans reflect and highlight its importance. The emergency plans now in place are directed to entire communities and reveal the importance of energy conservation. In responding to this challenge for energy conservation, one feature of the various schemes is an examination of on-going consumer trends and their effects on climate change. The problem with energy is that there is a progressive and inexorable rise in the cost of energy and an increasing demand on fossil fuel use. These two key indicators have led to a turning point for the importance of energy reduction, as well as the current legislative constrains on energy efficiency has been presented in Figure 4 (Phase I & II).

Couched within this emerging energy debate in the EU and its member states, the EU Framework Programme for Research and Innovation 2014–2020 includes in its action plan the need to legislate the policy priorities of the Europe 2020 strategy. This plan incorporates long-term aims for addressing the major concerns of energy demand shared by citizens in Europe and elsewhere. This strategy plan consists of different research areas related to the energy issue in the built environment. It is essential to determine that the objectives focus on social challenges such as: secure, clean and efficient energy; climate action, environment, resource efficiency, raw materials-secure, clean and efficient energy; and fighting and adapting to climate change [30].

The Directive 2012/27/EU indicates that the increasing level of dependence on energy imports and scarce energy resources brought into the EU and its member states sets unprecedented challenges for considering energy reduction. The climate change impact is also altered too. It has to be said that implementation of energy efficient technologies is an effective solution for addressing these challenges. Inevitably, the target plans assume to improve the EU's energy security in terms of reducing primary energy consumption and decreasing energy imports. Incidentally, implementation of energy-efficient technologies helps to reduce carbon dioxide emissions in a cost effective way. At the same time, these innovative technologies aim to mitigate climate change impact [30].

The current policy on energy-efficiency in Europe was adopted by the European Commission in March 2011. The Energy Efficiency Plan 2011 claims that the EU and its member states are responsible for operating a roadmap that will lead to an effective low carbon economy by 2050 [31]. For as this assessment indicates, the majority of action plans are undertaken for implementing energy-efficiency programmes across wide sectors of the economy but with particular emphasis on the residential sector. By considering carbon dioxide emissions, it is also noticeable that the Energy Efficiency Plan 2011 reports that the housing sector is estimated to be responsible for 41% of the total energy consumption in the EU [31]. It is also well understood that it is one of the most complex and articulated sectors to bring control mechanisms for implementing the energy-efficient technologies.

In particular, the T.R.N.C., because it is a small and isolated system as an island, depends primarily on imports to address its energy demands. The Republic of Cyprus energy system is isolated, resulting in significant barriers for the upgrade of the system and this creates high energy costs [32]. Furthermore, the Republic of Cyprus does not have any legislation and regulations concerning the energy performance of buildings; the energy saving potential was considerably high until becoming an EU member.

In this context, the research explores why the European Union Energy Performance Directive has assumed great importance and has become a very influential objective in our concern for the conservation of energy use in buildings since 2002. It is the first document to set the agenda for improving housing stock within the EU states and its purpose is to identify control mechanisms for the housing industry. To be able to set the agenda, there is a great deal of interest in retrofitting projects. It can be seen that it starts in the design-thinking process, through the construction phases, the redevelopment of buildings or regeneration of existing mass housing estate development regions within the association of concerning occupant's thermal comfort, to reduce energy consumption of buildings. Despite years of spuriously undertaken scientific research about conserving energy and the implementation of energy saving methods in the housing sector, the retrofit strategies have been on the board for at least a decade.

4.3. Energy Efficiency Programmes

Governing principles and implementation procedures that are useful in this context have arisen (Figure 6). Indeed, to achieve a reduction in energy consumption and an implementation of energy efficient technologies, on-going research in the field is needed [33]. According to above stated matters, this study uses the reveal of the energy demand in existing legislation from ELIH-Med (Energy Efficiency of Low Income Housing in the Mediterranean) and Era-co-build (European co-operative organisation) survey data, conducted in 2012 by the EU, which is the most recent data set available at the time of undertaking the research for this study. The definition of low zero carbon of several type of dwellings in Cyprus has been constituted by data collected from two on-going projects, "The Financing Mechanisms on Energy Efficiency of Low-Income Housing in Mediterranean Areas" (ELIH-Med) and "Countdown to Low Carbon Homes, ERACOBUILD" [34]. These projects are the first to be undertaken in Cyprus. The analysis has revealed some remarkable data concerning the low carbon emissions of selected pilot residential buildings in Cyprus. The ELIH-Med project aims to identify and implement innovative technical solutions to improve energy efficiency in low income housing in Cyprus. At the same time, the Era-co-build is aimed to create a larger market for systematic retrofitting schemes of residential buildings and investigated ways to motivate house owners, contractors and stakeholders to do an integral holistic retrofit. In general, the motivation and reasons for retrofitting of the existing housing stock are related to comfort improvement, quality of living and the necessity of optimising indoor thermal comfort conditions and improvements, and is also about energy saving targets. The results of ELIH-Med and Era-co-build data have been used to understand the changes in energy performance within the sector over the past decade [34]. Therefore, using the low-income housing retrofits to investigate the aims of this research was the most reasonable option at the starting point of this work.

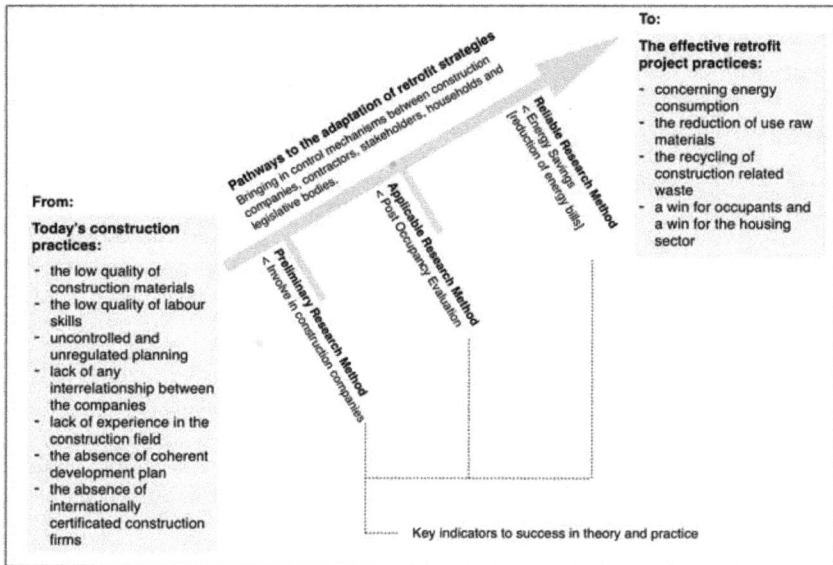

Figure 6. Model approach to the retrofit strategies.

Previous survey data for energy use patterns in the residential sector were focused primarily on individual technologies and did not consider occupant's thermal comfort and effective energy savings. However, this study exploits existing housing stock data for the T.R.N.C. to gain an insight into the key parameters related to energy use for applying energy efficient technologies. Traditionally, in the disciplines of architecture and building engineering, energy demand is addressed through individual building modelling based on the physical parameters of building form, fabric and systems. Simplistic and generalised assumptions are made about occupant's thermal comfort in terms of understanding occupant's energy use patterns and internal gains [35]. On the other hand, it is clear that counting carbon dioxide emissions of low-income housing through a conducted survey of ELIH-Med and Era-co-build addresses to what extent building energy performance, often assumed to relate primarily to physical building characteristics, is determined by interactions between occupant's thermal comfort and energy efficiency systems, as well as building and climate characteristics.

5. Method and Data

5.1. A Case Study Model: Northern Cyprus

The territory is characterised by a fragmented endlessly repetitive stream of self-built residential areas and privately-owned construction company built mass housing estates with no recognisable distinctions between city centres. Compared with other densely built Mediterranean cities, Northern Cyprus is dominated by large scale residential developments, not only in the coastal regions but also its mountainous regions, including urban agglomerations. At the same time, almost half (45%) of the owner-occupied building stock comprises self-built houses, often detached. The rest are low-rise flats (13%) or high-rise flats (44%) [35]. It is also noted that most urban agglomerations consist of a mix of housing types such as single and multi-family mix, single or multi-family and apartment block mix or apartment block and high-rise mix. Large scale residential tower blocks and mass housing estate developments, developed and regulated by privately owned construction companies. Such projects are often the size of whole city districts but are rarely geared towards the concept of a socially and functionally diverse and structurally open city.

Current problems are aggravated when considering the implementation of energy-efficient technologies that better suit occupant's thermal comfort needs are required, that is, mass housing estate developments. It is obviously seen that the supplied building stock could not match the energy efficiency implications and current tools to bring a linkage between existing construction practice and the power of adoption of energy design strategies for retrofitting existing mass housing estates in Northern Cyprus.

5.2. Rationale for Selecting Case Study Mass Housing Estate Developments

In response to research objective, and the need for a comprehensive, up-to-date analysis of case study buildings energy use, a "top-down" existing building stock analysis was applied. At the same time, a primary database was developed of annual energy use and associated building parameters for identification of energy use in three different construction eras in the 1970s, 1990s and 2010s in the residential sector. The research database was based on data collected under the Display Energy Certificate (DEC) scheme. A secondary research database was formed as a sub-set of revealing the monitoring results of 16 pilot-study buildings from the ELIH-Med (retrofitting of existing low-income buildings) research project in Cyprus. Statistical analysis was carried out to assess the impact of specific building parameters as energy determinants for both databases. To generalise findings and creating a representative sampling for the context, in accordance with research aim and objectives, an "archetype-based" method was taken. Archetype of three distinct construction periods (1970s, 1990s and 2010s) and quality of existing building stock were defined using data in the primary and secondary databases. Archetypes were based on three principal activity groups (demographic structure of households, building geometry and orientation) and two other forms of primary environmental strategy (heating and cooling demand) giving in three different retrofitting scenarios in total. These three concepts consider the housing stock in terms of characteristics, quality and developments. Distributions of results were obtained for each retrofitting options by analysing occupant's awareness in energy use and two principal forms relating to buildings' geographical location and climatic condition of the study area.

5.3. Research Design Model

This research consists of interdisciplinary collaboration in the area where single disciplinary studies often takes place. In that sense, there is communication and collaboration between research, design, and the implementation of policies and objectives for the construction industry. This research utilises a combination of qualitative research methods (ethnographic case study): on-site observations, semi-structured interviews and focus group discussions are all contained within this underlying approach.

Before undertaking these ethnographic studies, observations were carried out to include photographic documentation of housing estates, drawings, and maps of cities and housing estates. These observations are based on the collection of data relating to the selected case study mass housing developments in the field. After the collection of the necessary data through on-site observations, the study focuses on a case study approach to carry out analysis on the most problematic buildings in four different climatic regions of the research context. The researcher applied ethnographic studies as follows:

(1) Semi-structured interviews with construction company owners to understand the current condition of the construction industry and to understand the nature and benefits of implementing energy efficient technologies.

(2) Semi-structured interviews with house owners to understand their willingness to participate in implementing retrofit strategies in their homes.

(3)   Focus group discussions with house owners to investigate why house owners intend to be involved in the refurbishment activity of the recently built mass housing estates.

These methodologies were set out to address the issues of the housing sector. Although these research methods were tested in the T.R.N.C., it was designed to be applicable in the Republic of Cyprus with similar energy saving targets. In addition, the research hypothesis is that energy saving actions such as adoption of retrofit strategies could contribute to the reduction of the negative environmental impacts of the uncontrolled construction and refurbishment activities. This was mostly tested on mass housing estates, which were recently built, mainly by private construction companies.

This research includes some case studies, which consider different aspects of the housing estates such as location, characteristics, demographic structure of households and also information on the construction companies. The housing estates were chosen from 16 different projects in four different regions (coastal, inland, semi-mountainous and mountainous climatic zones) including urban and sub-urban areas, thus have a good representation of the common drivers in the property market with different levels of required refurbishment activity and different samplings from different climatic regions. It is further emphasised that the documentation of the field data has come from the identification of construction companies' projects, their policies, their targets and the problems they encountered in both the design and building process. This goes some way in providing information on the current condition of the industry in a particular centre or region.

During the research process, the researcher contacted 15 construction companies (Table 2). The research aim and targets were presented to get permission from these companies to examine their housing estate projects. For this purpose, 15 small and medium size architectural companies were identified in terms of their willingness to participate in the research process. These companies were key players, which have responded to the growing demand in the property market. Their structures and target groups showed variations within the location of the construction company and its projects. Before starting to conduct semi-structured interviews with households, a questionnaire-based survey on "Refurbishment Activity and Energy Consumption Patterns" was prepared.

This was partly to hear their views on how the retrofit strategies impacted on their own cultural assessment, but also to collect concrete examples of retrofitting experiences, which could be (anonymously) related to policy actors (institutions) to hear their responses. These data collection methods were to look at selected housing estate projects in terms of understanding typical energy consumption values and the effect of refurbishment activities. This method was also utilised in examining how occupants can play a key role during the implementation of the retrofit strategies. These interviews were intended to utilize information for each occupant's demographic structure needs and intentions in its involvement of any aspect of the refurbishment process.

Table 2. List of the interviewed construction companies.

No.	Company Name	Active Year	Location of Its Projects	Scale	Profile of the Company	Implementing of the Energy-Efficient Technologies	Target Group
1	Company A	1973	Famagusta-Iskele	Medium	Architecture firm in house expertise—holiday lets and commercial units	Low	Upper-middle income/middle income
2	Company B	1996	Famagusta	Medium	Architecture firm in house expertise	Non-used	High income/upper middle income
3	Company C	1989	Famagusta-Iskele	Medium	Architecture firm in house expertise—holiday lets and commercial units	Non-used	High income/upper middle income/middle income
4	Company D	1988	Famagusta-Iskele-Nicosia-Kyrenia	Medium	Architecture firm in house expertise	Non-used	High income/upper middle income
5	Company E	1984	Famagusta-Nicosia-Kyrenia	Medium	Architecture firm in house expertise and city planning	Strong	High income
6	Company F	1997	Famagusta-Iskele	Small	Architecture firm in house expertise and holiday lets	Non-used	Upper-middle income/middle income
7	Company G	1988	Kyrenia	Small	Architecture firm in house expertise	Non-used	Upper-middle income/middle income
8	Company H	1995	Famagusta-Kyrenia	Small	Architecture firm in house expertise	Non-used	High income/upper middle income
9	Company J	1991	Kyrenia	Medium	Architecture firm in house expertise	Non-used	High income/upper middle income
10	Company K [funded by Turkey]	1984	Famagusta-Nicosia-Kyrenia	Medium	Architecture firm in house expertise—commercial units—urban planning	Non-used	Middle income/low income
11	Company L	1980	Famagusta	Small	Architecture firm in house expertise and commercial units	Non-used	Middle income
12	Company M	2003	Famagusta-Nicosia-Kyrenia	Small	Architecture firm in house expertise-commercial units and infrastructure	Non-used	Upper-middle income/middle income
13	Company N	2003	Famagusta-Iskele	Small	Architecture firm in house expertise—holiday lets and commercial units	Non-used	High income/upper middle income/middle income
14	Company P	1995	Kyrenia	Medium	Architecture firm in house expertise and tourism developments	Non-used	High income
15	Company R	2003	Nicosia	Small	Architecture firm in house expertise	Non-used	High income/upper middle income

The people who were interviewed in order to participate in this study are all residents of single- or multi-family owner occupied housing units. The participants were people who do not identify themselves as vulnerable. Each was given a questionnaire to complete and was also interviewed by the researcher. Furthermore, the economic, physical, social and cultural environment in which the study was guided was observed. This approach combines regular site visits to the same households for two seasons (summer and winter) for a report on the environmental impact of the built environment over a period of one year in different climatic locations. The researcher contacted households in different project sites to get permission to re-undertake the questionnaire survey in the following research period. The interview guide was therefore pilot-tested. The objective of this methodology is to calibrate the policies of implementing the adaptation of retrofit strategies to illustrate a trend of refurbishment activity in the recently built mass housing estates. This method was used as background information for this study to fill a research gap and contribute to knowledge on implementing an adaptation of retrofit strategies.

The qualitative analysis software of NVivo (QSR International: Melbourne, Australia) was used to analyse the fieldwork data. The analysis was guided by a preliminary thematic analysis of the key concepts prompted on the interviews. The first three concepts consider the housing stock in terms of characteristics, quality and developments. The other three concepts deal with current polices and action plans to introduce control mechanisms for retrofitting projects. It should also be noted that the semi-structured interviews and focus group discussions were conducted with households only on certain selected buildings and so the findings of the study apply to the narrow field under investigation and was not broadened to include some form of generalised opinion.

Another contribution to the field is the general evaluation carried out on understanding energy consumption of recently built housing estates in the T.R.N.C. As this study shows, there is very little research available or undertaken in the academic world that targets "retrofit strategies". This research concept finally led to assessing and generating new pathways of research and innovative design tools in the management of the mass housing renewal and urban development but at the same time involving the notion of a socio-cultural paradox. The aspect of this research is to understand how it is best possible to integrate the application of energy efficiency technologies to the adaptability of the prototype retrofit scenario. It included cross-cultural studies as a research concept to investigate the pattern interpretation of energy consumption use and retrofitting.

5.4. Undertaken Approach for Conducting on Questionnaire Survey

To ensure systematic analysis of the key aim and objectives of the research, the methodology adopted particularly for the mass housing estate developments is hereby explained. The design chosen for this research is a "Before-and-after design", which, as the name implies, is a set of semi-structured interviews with house owners and privately-owned construction companies taken from a group of respondents, who are then subjected to an experimental variable before being undertaken again. In this case, a two-phased questionnaire survey is conducted (phase A and phase B) where phase A is used as a control group. The experimental variable is the articulated refurbishment work done by respondents. The sample of respondents are not the same; however, they are matched samples in that they both are within four climatic regions in the T.R.N.C., are mass housing estate development projects, have been regulated privately owned construction companies, comprise similar sized, solid walled system without any insulation material implementation and energy-inefficient houses, and all are common representative residential building typology for the undertaken research context. The socio-demographic analysis undertaken later in the survey analysis ensures significantly comparable samples of respondents in both phases.

An analytical survey was administrated to identify how households of selected mass housing estate buildings identified to be energy-inefficient react to and perceive the challenges of implementation of energy efficiency building systems. The survey was also conducted to develop an understanding of house-owners' behaviour during the un-regulated refurbishment process.

The questionnaire-based survey therefore investigates occupants' and privately-owned construction companies' experience during the energy upgrade works are done by means of a two-phased survey questionnaire (A and B).

According to the State Planning Organisation, Economic and Social Indicators statistics (2015), The Follow up and Coordination Department in the T.R.N.C. comprises 3246 households, of which 1200 private and social houses were eligible for the sampling criteria [36]. The sample fraction initially aimed for was 10% of the total selected households; however, 70 households were successfully recruited for the study, constituting 4% of the total, which is still relatively reasonable fraction. The researcher decided to use a stratified sampling approach to select households from the State Planning Organisation statistical index of representative mass housing estate development projects eligible for the study. The questionnaire includes ethnographic background information about the respondents and the household, a set of questions about environmental attitudes and values, and questions about behaviour related to the building. More specifically, some questions are based upon the amount of energy advice and information provided to households for understanding energy conservation measures. A few questions were prepared, including:

- Have you made any changes since your house/apartment you bought it? If "Yes", what sort of changes did you make?
- At this stage would you consider ways of reducing energy consumption?
- Do you know anything about energy saving methods?
- What do you think would be the benefits of using energy-efficient technologies?
- Do you know anything about the energy efficiency objectives of the European Union?
- Do you know how implementing effective retrofit strategies can reduce annual energy bills?

6. Analysis and Results

The study found that refurbishment activities are identified according to the degree to which the building systems used by occupants with reference to three main indicators are interrelated: the age of the building, the construction material and its energy demand. Through these variations, it is possible to define a decay representing all major classes of buildings in the residential sector and to utilise obsolescence as an indicator in an analysis of the buildings [36]. This is one of the main reasons why different building typologies are widely investigated and considered strategic in the selected mass housing estates. For these reasons, an expected analysis is co-related to both these parameters concerning the construction material and its system, the obsolescent part of the building systems, the energy efficient requirements and the interventions that come into play over time.

In Figure 7, it can be seen that more than half the sample has already replaced their building systems, while 42% install energy efficient lightings. This may be due to the high numbers of electricity suppliers that provide energy efficient light bulbs and economically inactive respondents who might spend most of their time at home, thus consuming more energy than households with employed members.

Another important fact is the role of the house owners' requirements in the design process and how these may impact the construction process. Therefore, the great challenge is to create collaborative mechanisms whereby both the construction companies and the house owners may contribute and coordinate efforts in solving the energy problem. To reach this goal, the demographic structure of occupants and their behaviour is required. Each single household has to be convinced and assured of the reasonableness and the economic advantages that will accrue from investing in the improvement of building systems concerning energy efficiency in and on their property. Thus, to obtain reliable and effective results in terms of energy efficient improvements, it is important that the interventions are articulated by occupants in the housing sector, as they represent now the common problems of the new-built housing stock. In this research context, the selected mass housing estate development projects are composed of heterogeneous buildings with different typologies and dimensions, and consequently with different built purposes applied.

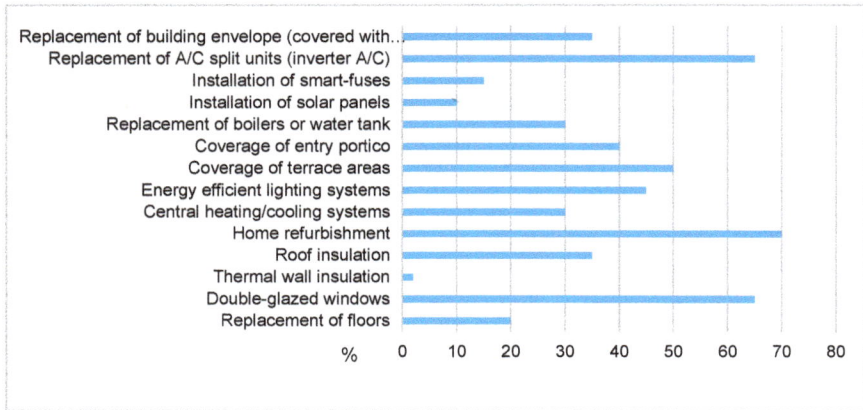

Figure 7. Refurbishment activity patterns.

Today, primary needs have evolved according to the demographic structure of the occupants and their lifestyles, financial capacities and several different occupants' profiles are identified naturally. These householder profiles differ according to which housing estate project is under discussion and what the construction companies' targets were during the design process. However, it should also be remembered that house owners are now much more willing to improve the physical conditions of their dwellings than they were in the past and the trend for refurbishment continues. Consequently, profiling occupants and their behaviour is a much more complex activity when it also considers the refurbishment habits of the occupants. Construction systems and materials are common nominators in the selected buildings and are strictly connected with the choice of construction companies' progress at the time. Therefore, the materiality of buildings is a significant indicator not only of its level of physical obsolescence, but also the rapid construction demand linked to implement poorly built materials at that time.

Several differences occur among apartments, terraced houses, semi-detached and detached buildings due to typology, built-form, distribution and construction systems. The common factor is generally that the housing stock after the great expectations raised by the Annan Plan was not designed to meet today's energy efficient standards especially concerning the control of indoor comfort conditions (heating and cooling demand), and the thermal losses due to poorly-built construction material choices. This means that offering adaptation of retrofit packages for improving energy efficiency in the housing sector would not only bring a relevant reduction in energy consumption in the selected housing estates, but also that households can be strongly involved in reducing their energy costs.

In response to questions concerned with occupants' energy awareness and behaviour in energy use, 60% are aware of challenges of installation of energy efficiency building systems as can be seen in Figure 8. This may be due to the consistent information campaigns communicated via media sources about energy conservation and recycling. The State Planning Organisation statistics in 2015 reported that energy awareness and recycling rates for households have risen approximately 40% in the T.R.N.C. from the previous year rate [36].

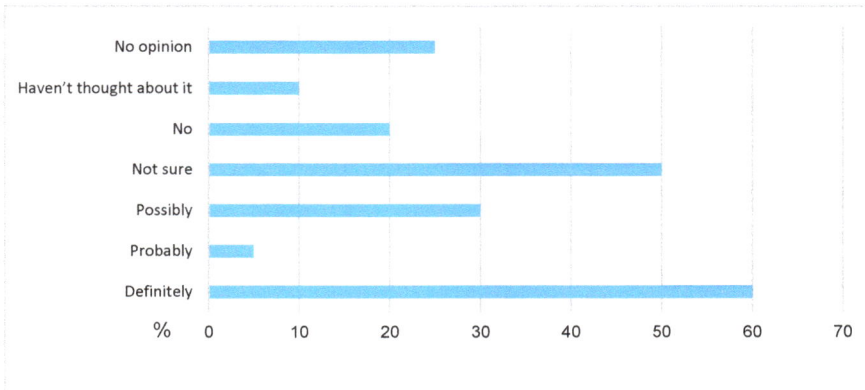

Figure 8. Energy conservation awareness.

In addition, the study also revealed that changes of building components, as articulated by occupants, are on the rise. Even though some buildings have undergone major renovation, the addition of more spaces and the covering of terrace areas account for a high percentage of the renovation activity. Hence, the quality profile of this new-built housing stock changes gradually and only the construction process and renovations differ substantially from the housing stock.

Subsequently, it is quite hard to have a reliable forecast of the renovation trends without having a control mechanism and adaptation packages developed through an investigation of the selected buildings. At this stage, the preliminary findings show that identification of the diagnosis in selected buildings can be useful for understanding what main changes occupants are expecting in the residential sector, and for further investigating their requirements from the construction companies and their involvement in the decision-making process.

Moreover, the study found that the occupant's refurbishment trends affect not only the energy performance of buildings but also increases carbon dioxide emissions in the environment. It is also worth commenting that the impact of the construction activity has produced further problems in the residential sector and this contributes to making an investigation of the potential adaptation packages much more complex. To compensate for this, a useful starting point could be identified as the diagnosis of buildings in the selected mass housing estates. What seems to be clear is that the physical quality of the buildings, the demographic structure of households and the quality of refurbishment activity are perceived as inappropriate in meeting the emerging demand of the residential sector.

7. Discussion

In this case study approach, 70 buildings are analysed, 70 semi-structure interviews with house-owners are conducted, and five retrofitting strategies are considered (Table 3), in accordance with the contribution they make to reducing energy demand.

In these selected housing estates, three main requirements have been associated with refurbishment trends. The first one deals with the covering of terrace spaces in the detached and semi-detached buildings and balconies in the apartments and terraced houses and adding more room spaces as a whole (which is strictly related to the demographic structure of the occupants) to obtain two different kinds of result. First, adaptable spaces allow for the extension of the living room, dining area and entry lobby of the buildings according to changes in the lifestyles of occupants. Second, it allows some spaces to be widened and given a more specialised function, such as ample living spaces for a family or a room. This could be achieved by extending the spatial layout of the existing building. These kinds of interventions are generally integrated to the on-going changes of users' profiles and to the trends of refurbishment activities and do not belong to concerns about reducing energy consumption demand but rather to the idea of improving the quality of living conditions.

Table 3. Structure of the step-by-step applicable "retrofit strategies".

STEPS	RETROFIT STRATEGIES
S1	Replacement of existing windows
S1B	Integration of replacement of existing equipment, heating/cooling system
S1C	Integration of PV and solar collectors on the roof
S2	Building envelope implementation of roof and partially of facades to avoid thermal bridges
S3	Total building envelope implementation
S4 (S1 + S2)	Replacement of existing windows, total building envelope implementation (optional integration or replacement of existing equipment, heating/cooling system)
S4B	Integration of PV and solar collectors on the roof/facades
S5	Volumetric additions, partial replacement of existing windows, partial building envelope implementation (optional integration or replacement of existing equipment, heating/cooling system)
S5	Integration of PV and solar collectors on the roof/facades

The second refurbishment activity involves a general improvement of the dwelling and also of the building as a whole in terms of the replacement of kitchen units or bathrooms, roof insulation, installation of double-glazing windows and addition of shading panels (pergolas) directly to the outside of the building. Most of these activities are an expression of trends of informed high-quality interventions. Therefore, the problem is related to understanding the benefits of "energy-efficiency" during the refurbishment process to meet the requirements of building standards. In most cases, the refurbishment activity is perceived by the occupants as not only improving the quality of living conditions but also a real opportunity to reduce energy consumption in the residential sector. The third refurbishment activity deals with access to fresh water supply, the recycling of rain water and grey water and the connection to the grid which varies depending on the location of each housing estates. Nevertheless, these activities to improve the infrastructure of the buildings may have relevant effect in terms of the utilisation level of the mass housing estates. At the same time, many buildings are being fitted with solar panels.

In Figure 9, with regards to "reasons for taking any of the retrofitting scenarios into an action", saving money came first as a matter of concern where the majority of both samples reported this to be the main reason, along with one or more other reasons such as saving energy, due to habit, and environmental concern. Around a quarter of the samples took these actions only to save money as the one of the main reason, followed by to save energy and due to environmental concern, while around a fifth take these actions out of habit. Considering the f first concern for most households in this study is to save money, it suggests that financial incentives could possibly be effective in encouraging policy uptake and delivery in this particular region.

Energy retrofit for recently-built housing stock: The way forward foreseen here is to put in place effective implementation scenario(s) which are decided upon after the buildings have been undergone a systematic retrofit. These strategies aim to reduce energy consumption for the heating and cooling demand of buildings, as well as improving the indoor comfort conditions of buildings by encouraging the benefits of using energy efficiency technologies in maintenance interventions and reducing the up-front costs of the existing condition of building systems before going under refurbishment. From this study, it was found that the design of retrofitting strategies, while allowing higher residual thermal loads from individually refurbished buildings compared to a standard renovation, reaches 91%, 92% and 87% lower primary energy balance than the existing residential building stock [37].

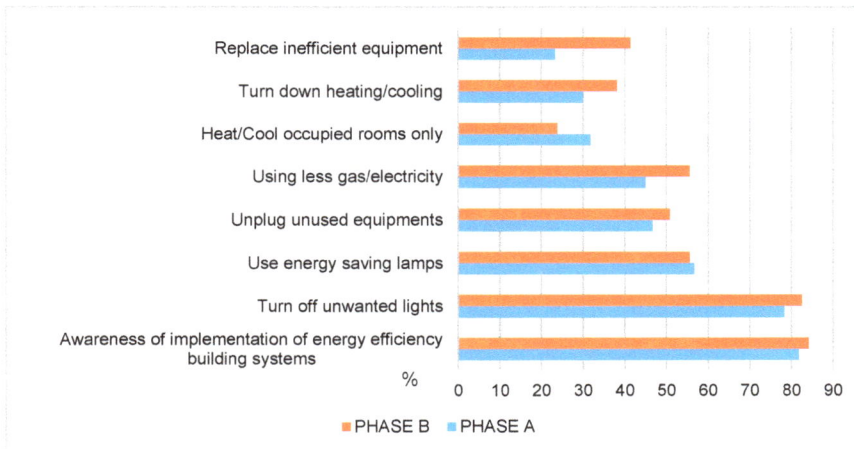

Figure 9. Behaviour patterns in phases A and B.

There is an increase in demand on the property market for energy efficient designed buildings focusing on energy consumption patterns of households with regard to environmental impact as a result of poorly built construction materials and climate change impact. Therefore, a concern for the life-cycle performance of buildings is a central purpose in any refurbishment activity. An analysis of the life cycle of buildings is a pivotal factor in any measurement of performance when assessing the impact on the environment from buildings of any kind. This approach is more feasible to implement, because of the energy efficient alternative when the construction systems have reached the end of its life length and need refurbishment either way, or to advance measures that were not yet due, in order to improve energy efficiency [38].

Under this rendering of the implementation of energy-efficiency technologies lies a further level of case study analysis. This focuses attention on the diagnosis, action and intervention of construction companies, house owners, building contractors, stakeholders, architects and designers in the T.R.N.C. as well as those interested in furthering the promotion of effective retrofitting scenarios with a view to reducing the energy consumption of buildings. By considering the residential sector, the importance of the financial viability of such privately owned construction companies cannot be underestimated. For a large-scale construction company to have an impact, it must be successful and financially viable [39].

It has also to be said that the most common refurbishment activity is the addition of new volumes adjacent to the existing building or open terrace areas on both the ground and first floors. These often unintentionally lead to increase in energy consumption of the buildings. Consequently, the changes on the current market conditions after the Annan Plan lead to the residential sector to becoming aware of the necessary EU objectives in the design process as regards to energy efficiency and also in defining the construction process and its impacts on the built environment.

Therefore, based on the arguments discussed above, it is important to incorporate in the design a process outline of "energy-efficiency", such as retrofitting strategies so that it will have a high market appeal and therefore should be incorporated into the construction phase so that the annual consumption of energy can be reduced as much as possible. It should respond to growing demand of construction activity. For this reason and in the process of the implementation of retrofitting strategies for reducing the energy consumption of buildings, it is apparent that the role of government initiatives, the framework policies of construction companies and house owners should be examined as a way of finding practical routes to implement "retrofit strategies" and purpose design recommendations for the housing sector in the future (Figure 10).

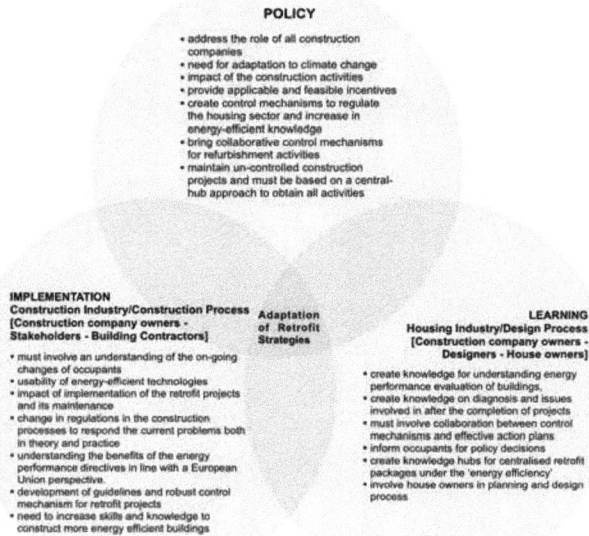

POLICY
- address the role of all construction companies
- need for adaptation to climate change
- impact of the construction activities
- provide applicable and feasible incentives
- create control mechanisms to regulate the housing sector and increase in energy-efficient knowledge
- bring collaborative control mechanisms for refurbishment activities
- maintain un-controlled construction projects and must be based on a central-hub approach to obtain all activities

IMPLEMENTATION
Construction Industry/Construction Process
[Construction company owners -
Stakeholders - Building Contractors]
- must involve an understanding of the on-going changes of occupants
- usability of energy-efficient technologies
- impact of implementation of the retrofit projects and its maintenance
- change in regulations in the construction processes to respond the current problems both in theory and practice
- understanding the benefits of the energy performance directives in line with a European Union perspective
- development of guidelines and robust control mechanism for retrofit projects
- need to increase skills and knowledge to construct more energy efficient buildings

Adaptation of Retrofit Strategies

LEARNING
Housing Industry/Design Process
[Construction company owners -
Designers - House owners]
- create knowledge for understanding energy performance evaluation of buildings,
- create knowledge on diagnosis and issues involved in after the completion of projects
- must involve collaboration between control mechanisms and effective action plans
- inform occupants for policy decisions
- create knowledge hubs for centralised retrofit packages under the 'energy efficiency'
- involve house owners in planning and design process

Figure 10. The hypothesised relationship between policy, implementation and learning through research process.

8. Conclusions and Recommendations for Future Research

8.1. Conclusions

From the findings of the study, it can be concluded that the original contribution of this research lies in adoption of the retrofit strategies that systematically integrates energy efficiency standards in order to improve conditions within the residential sector under the combined influence of three variables, namely the construction activities, occupants' behaviour and the energy consumption of buildings. Nevertheless, the majority of households (54%) indicate that they would use energy efficient technologies for their home refurbishment to upgrade energy performance. At the same time, 40% of the households would not implement cost-effective energy efficiency building systems and 6% was not aware of challenges through "systematic" retrofitting. In summary, this study and investigation process analyses the pattern interpretation of the occupants' behaviour and their cultural assessment embedded energy performance of buildings during the implementation of retrofit strategies. In this context, no existing research was identified applying energy efficiency standards of retrofitting to any types of buildings, whether recently built mass housing estate projects or otherwise.

The results of the research contributed to facilitate private construction companies that aim to support retrofit strategies by providing them with new guidelines and policies together with the necessary data about the implementations needed for the improvement of the housing sector in the T.R.N.C. This research does not only enable households to become involved in the process of identifying the applicable retrofit scenarios, but also to improve living conditions and to achieve minimum energy efficiency consumption of buildings. This research has provided the context with the knowledge to realise the main goals: the identification of possible future instruments and incentives that are needed to overcome the weakness in current legislation and thus bring about a more energy efficiency residential building sector.

8.2. Recommendations for Future Research

Retrofitting and upgrading of the existing mass housing estate developments to energy efficiency through testing different retrofitting strategies is of utmost importance but their long-term viability depends on the sustainable, holistic approach where energy related measures are closely linked with functional, construction, and economic demands. The implementation of the preliminary architectural and energy efficiency improvements in the residential buildings would have the additional benefits of increasing the housing space market value and positive social effects, and incentivising house owners' awareness of energy consumption. At the same time, energy efficiency implementations targeted at the selected case study buildings can also improve indoor air and indoor environmental quality, with corresponding to reducing overheating risk assessment of a building. In addition to direct energy consumption reduction, energy use potentialities of retrofitting measures can provide indirect economic benefits to both house owners and privately owned construction companies, particularly in this research context.

Introducing subsidies/implications improves the feasibility of undertaken energy conscious retrofitting strategies, but they become sustainable on their own aspects and design parameters. Undertaken energy performance analysis of prototype models shows clearly that the proposed energy efficient strategies must be feasible. Obtained results must be valid and reliable justification of the significance of the proposed redevelopment project in an institutional level to bring significant potentialities for energy savings. To conclude, optimising energy efficient retrofitting measures for mass and large scale residential developments are of crucial importance, thus the existing housing stock can be treated in a systematic manner to achieve similar energy saving potentials.

Author Contributions: Bertug Ozarisoy and Hasim Altan conceived and designed the concept and outline for the paper; Bertug Ozarisoy conducted the interviews and wrote the paper; and Hasim Altan supervised, and provided sources, comments, and major edits to the paper.

Conflicts of Interest: The authors declare no conflict of interest.

References

1.   United Nations. *World Urbanisation Prospects: The 2007 Revision*; Department of Economic and Social Affairs, United Nations: New York, NY, USA, 2008; Available online: http://www.un.org/esa/population/publications/wup2007/2007WUP_Highlights_web.pdf (accessed on 15 February 2008).

2.   Intergovernmental Panel on Climate Change (IPCC). *Climate Change: The Physical Science Basis*; Contribution of Working Group I to the Fourth Assessment Report of the Intergovernmental Panel on Climate Change; Solomon, S., Qin, D., Manning, M., Chen, Z., Marquis, M., Averty, K.B., Ignor, M., Miller, H.L., Eds.; Cambridge University Press: Cambridge, UK, 2007.

3.   Ministry of Environment and Natural Resources Department of Meteorology-TRNC (Cevre ve Dogal Kaynaklar Bakanligi Meteoroloji Dairesi Mudurlugu in Turkish). *The Annual Report, Climate of Cyprus*; Department of Meteorology: Nicosia, Turkish Republic of Northern Cyprus, 2015.

4.   Ghosh, B.N.; Aker, S.L. Future of North Cyprus: An economic-strategic appraisal. *Futures* **2006**, *38*, 1089–1102. [CrossRef]

5.   State Planning Organisation—TRNC (Devlet Planlama Orgutu—KKTC). Macroeconomic Developments, Main Objectives and Macroeconomic Targets of 2008 Programme. pp. 105–138. Available online: http://www.devplan.org/Frame-eng.html (accessed on 14 August 2012).

6.   Cyprus Meteorological Service. *Meteorological Statistical Data for Cyprus: The Annual Report*; Department of Meteorology: Nicosia, Turkish Republic of Northern Cyprus, 2013.

7.   Cohen, B. Urbanisation in developing countries: Current trends, future projections, and key challenges for sustainability. *Technol. Soc.* **2006**, *28*, 63–80. [CrossRef]

8.   Ulucay, P. The European Spatial Planning Approach: An Instrument towards the Creation of a Sustainable Housing Policy in Northern Cyprus. Ph.D. Thesis, Housing, Joint Symposium by HERA-C and HREC. Eastern Mediterranean University, Gazimagusa, Turkish Republic of Northern Cyprus, 2008.

9.   Ulucay, P. A Critical Evaluation of the Town Planning Law of Northern Cyprus in line with the European Spatial Development Perspective. Ph.D. Thesis, Eastern Mediterranean University, Famagusta, Turkish Republic of Northern Cyprus, 2013, unpublished.

10. Ratti, C.; Baker, N.; Steemers, K. Energy consumption and urban texture. *Energy Build.* **2005**, *37*, 762–776. [CrossRef]

11. Swan, L.; Ugursal, G. Modelling of end-use energy consumption in the residential sector: A review of modeling techniques. *Renew. Sustain. Energy Rev.* **2009**, *13*, 1819–1835. [CrossRef]

12. Yorucu, V.; Keles, R. The Construction Boom and Environmental Protection in Northern Cyprus as a Consequence of the Annan Plan. *Constr. Manag. Econ.* **2007**, *25*, 77–86. [CrossRef]

13. Mehmet, O.; Yorucu, V. Explosive construction in a microstate: Environment limit and the Bon-curve: Evidence from North Cyprus. *Constr. Manag. Econ.* **2008**, *26*, 79–88. [CrossRef]

14. Safakli, O. An overview of the construction sector in Northern Cyprus. *Afr. J. Bus. Manag.* **2011**, *5*, 13383–13387.

15. Balkiz, Y.; Therese, W.-L. Small but Complex: The Construction Industry in North Cyprus. *Procedia Soc. Behav. Sci.* **2014**, *119*, 466–474. [CrossRef]

16. Bourdic, L.; Salat, S.; Nowacki, C. Assessing cities: A new system of cross-scale spatial indicators. *Build. Res. Inf.* **2012**, *40*, 592–605. [CrossRef]

17. Papadopoulos, A.M.; Oxizidis, S.; Papanritsas, G. Energy, economic and environmental performance of heating systems in Greek buildings. *Energy Build.* **2008**, *40*, 224–230. [CrossRef]

18. Electricity Authority of Cyprus—TRNC (Kibris Turk Elektrik Kurumu—KKTC). *The Annual Report, Energy Consumption of Residential Buildings*; Ministry of Environment and Natural Resources, Department of Energy: Nicosia, Turkish Republic of Northern Cyprus, 2015.

19. Lechtenbohmer, S. Compliance with Building Regulations. Presented at IEA 'International Workshop on Meeting Energy Efficiency Goals', Enhanced Compliance, Monitoring and Evaluation, Stream 1: Buildings, Paris, France, 28–29 February 2008.

20. Beerepoot, M. *Energy Policy Instruments and Technical Change in the Residential Building Sector*; IOS Press: Amsterdam, The Netherlands, 2007.

21. Itard, L.; Meijer, F. *Towards a Sustainable Northern European Housing Stock*; IOS Press: Amsterdam, The Netherlands, 2008.

22. Sunikka, M.M. *Policies for Improving Energy Efficiency in the European Housing Stock*; IOS Press: Amsterdam, The Netherlands, 2006.

23. Itard, L.; Meijer, F.; Vrins, E.; Hoiting, H. *Building Renovation and Modernisation in Europe State of the Art Review*; OTB Research Institute, Delft University of Technology: Delft, The Netherlands, 2008.

24. Engelund Thomsen, K.; Wittchen, K.B. *European National Strategies to Move towards Very Low Energy Buildings*; Danish Building Research Institute, Aalborg University: Aalborg, Denmark, 2008. Available online: http://sbi.dk/Assets/European-national-strategies-to-move-towards-very-low-energy-buildings/2008-03-13-3730310829.pdf (accessed on 22 September 2011).

25. Schule, R. *Energy Efficiency Watch Final Report on the Evaluation of National Energy Efficiency Action Plans*; Wuppertal Institute: Wuppertal/Berlin, Germany; Ecofys Germany: Cologne/Berlin, Germany, 2009. [CrossRef]

26. Hamiltion, B. *A Comparison of Energy Efficiency Programmes for Existing Homes in Eleven Countries*; Vermont Energy Investment Corporation: Burlington, VT, USA, 2010. Available online: https://www.ucalgary.ca/tsenkova/files/tsenkova/2-CHAPTER01NIEBOERETAL.pdf (accessed on 31 August 2011).

27. Fay, R.; Treloar, G.; Iyer-Raniga, U. Life-cycle energy analysis of buildings: A case study. *Build. Res. Inf.* **2000**, *28*, 31–41. [CrossRef]

28. Forsberg, A.; Von Malmborg, F. Tools for environmental assessment of the built environment. *Build. Environ.* **2004**, *39*, 223–228. [CrossRef]

29. Itard, L.; Klunder, G. Comparing environmental impacts of renovated housing stock with new construction. *Build. Res. Inf.* **2007**, *35*, 252–267. [CrossRef]

30. EUR-Lex. Official Journal of the European Union, C Series. *Off. EU Lang.* **2012**, *24*, 134.

31. Cyprus Energy Agency (CEA). Investigation of the Different Regulatory Frameworks Regarding Territorial, Landscape and Energy Planning in Each Partner's Region. ENERSCAPES. Available online: http://ftz.org.mt/wpdemo/wordpress/wp-content/uploads/2014/06/enerscapes_regualtory_frameworkfinalised.pdf (accessed on 2 June 2015).

32. European Commission. Directive 2010/31/EU of the European Parliament of the Council of 19 May 2010 on the energy performance of Buildings (recast). *Off. J. Eur. Union* **2010**. Available online: http://www.buildup.eu/en/practices/publications/directive-201031eu-energy-performance-buildings-recast-19-may-2010 (accessed on 4 March 2011).

33. Panayiotou, G.P.; Kalogirou, G.A.; Florides, C.N.; Maxoulis, A.M.; Papadopoulos, M.; Neophytou, P.; Fokaides, G.; Georgiou, A.; Symeou, G. The characteristics and the energy behavior of the residential building stock of Cyprus in view of Directive 2002/91/EC. *Energy Build.* **2010**, *42*, 2083–2089. [CrossRef]

34. Panayiotou, G.P.; Charalambous, A.; Vlachos, S.; Kyriacou, E.; Theofanous, E.; Filippou, T. *Increase of Energy Efficiency of 25 Low-Income Households in Cyprus*; Department of Mechanical Engineering and Materials Science and Engineering, Cyprus University of Technology: Limassol, Cyprus, 2012.

35. The SouthZEB Report. *nZEB Training in the Southern EU Countries. Maintaining Buildings Traditions*; Report on the Current Situation Regarding nZEB in the Participating Countries; Report No. WP2-Deliverable 2; The Intelligent Energy Europe Programme of the European Union; European Commission: Brussels, Belgium, 2014.

36. State Planning Organisation—TRNC (Devlet Planlama Orgutu—KKTC). Economic and Social Indicators Statistics. 2015. Available online: http://www.devplan.org/Frame-eng.html (accessed on 8 August 2017).

37. Conci, M.; Schneider, J. A District Approach to Building Renovation for the Integral Energy Redevelopment of Existing Residential Areas. *Sustainability* **2017**, *9*, 47. [CrossRef]

38. Hogberg, L.; Lind, H.; Grange, K. Incentives for Improving Energy Efficiency When Renovating Large-Scale Housing Estates: A Case Study of the Swedish Million Homes Programme. *Sustainability* **2009**, *1*, 1349–1365. [CrossRef]

39. Saintier, S. Community Energy Companies in the UK: A Potential Model for Sustainable Development in 'Local' Energy? *Sustainability* **2017**, *9*, 1325. [CrossRef]

sustainability

MDPI

Article

Overheating and Daylighting; Assessment Tool in Early Design of London's High-Rise Residential Buildings

Bachir Nebia [1],*and Kheira Tabet Aoul [2]

[1] Roberts and Treguer Ltd., London E1 7SA, UK
[2] Architectural Engineering Department, United Arab Emirates University, P.O. Box 15551 Al Ain, UAE; kheira.anissa@uaeu.ac.ae
* Correspondence: bachir.nebia@gmail.com; Tel.: +44-781-870-3036

Received: 30 June 2017; Accepted: 23 August 2017; Published: 30 August 2017

Abstract: High-rise residential buildings in dense cities, such as London, are a common response to housing shortage. The apartments in these buildings may experience different levels of thermal and visual comfort, depending on their orientation and floor level. This paper aims to develop simplified tools to predict internal temperatures and daylighting levels, and propose a tool to quickly assess overheating risk and daylight performance in London's high-rise residential buildings. Single- and double-sided apartments in a high-rise building were compared, and the impact of their floor level, glazing ratio, thermal mass, ventilation strategy and orientation was investigated. Using Integrated Environmental Solutions Virtual Environment (IES VE), temperature and daylight factor results of each design variable were used to develop early design tools to predict and assess overheating risks and daylighting levels. The results indicate that apartments that are more exposed to solar radiations, through either orientation or floor level, are more susceptible to overheat in the summer while exceeding the daylighting recommendations. Different design strategies at different levels and orientations are subsequently discussed.

Keywords: overheating; daylighting; design tool; assessment tool; London; high-rise; residential; floor-level; orientation; glazing

1. Introduction

The planet is changing. Average global temperatures are expected to increase 5 °C by 2100 [1] leading to increased frequency and intensity of summer heatwaves in urban settings. The consequences could be more devastating than the European heatwave of August 2003, which led to 70,000 deaths [2], including 2000 in the United Kingdom [3]. Due to climate change, the number of deaths from excessive heat could be three times higher by 2050 [4] as the temperatures experienced during the summer of 2003 might become the norm by the 2040s [5].

In urban environments, the heat island effect will exacerbate this phenomenon [6,7]. Cities and buildings will struggle to exhaust the heat and cool down at night. Furthermore, by 2050, 66% of the world's population is expected to live in cities [8]. In vulnerable urban environments such as London's, where a 9 °C temperature difference was observed between the inner-city and the surrounding areas during the night-time of the 2003 heatwave [9], heat risk emerges as a significant climate change issue. As a result, multiple policies aiming to reduce energy use and carbon emission emerged. For example, the UK Committee on Climate Change (CCC) implemented a 20% reduction target on the space heating demand [10], resulting in an increased level of insulation in new buildings. This led to thermal improvements in winter [11,12] and a reduction in energy consumption for new and refurbished dwellings. Policies of this type have triggered the development of several design guidelines and energy performance standards such as the ones published by the Zero Carbon Hub and the PassivHaus Trust. However, poorly implemented energy efficient design strategies have shown their limits and negative impacts [13]. For instance, an excessive internal heat due to highly insulated and airtight construction without appropriate passive cooling design strategies could compromise occupants' health and comfort in residential buildings [14]. In the UK, 20% of households may already be at risk of overheating [15], and, with the predicted climate change, this percentage will likely increase [4]. Although no universal definition of overheating exists, the phenomenon has been widely monitored [15–24], and thermally modelled [25–30] using either static or adaptive assessment criteria. Numerous studies have examined a number of dwelling types that represent broadly the housing stock in the UK. However, it is difficult to compare the results from the different studies because of a lack of standardization in their input parameters [31].

As an example of these modelling studies, series of early researches [32–34] have looked at thermal comfort of four different residential types: detached house, semi-detached house, town house and a top floor flat. The results indicated that an increase in thermal mass could significantly mitigate the overheating issue and that careful decisions should be made with regards to the design parameters that control solar heat gains. The results confirmed that top-floor apartments are the most exposed to the overheating risk in tower blocks. Furthermore, Coley and Kershaw [35] investigated a large number of buildings by testing their thermal response to 400 different variables of four build-forms (a house, a purpose-built flat, an office and a school). A linear relationship between the weather and the internal thermal conditions was found while climate change will further increase the internal mean temperatures. However, these studies were mainly based on dynamic thermal simulation methods and are exposed to a certain number of assumptions and uncertainties [30].

In another survey of dwelling types that are at a higher risk of overheating [36], 185 cases were reported in two consecutive surveys with 73% of them located in urban areas. Two thirds (73%) of observed cases were apartments of which around 37% were located on the top floors. It was reported that 48% of the overheating cases were new flats. While the proportion of new flats is small compared to the total housing stock [37], the large number of overheating cases reported in this study suggests a considerable risk of thermal discomfort in newly-built flats at higher floors; this agrees with the study carried out by Vandertorren et al. [38]. Since overheating instances were reported by residents, the difference between the measured overheating and the perceived could be significant depending on several factors such as the occupants' demographics and their level of activity. This study could be considered as "the tip of the iceberg" [36]; however, the lack of data and reporting process is a significant barrier to understanding the extent of overheating in the UK and consequently to the design of effective solutions.

Both monitoring and modelling studies have their limitations. In response, a recent study attempted to reconcile both monitoring and modelling approaches [22]. A statistical meta-model was created from over 3400 building combinations and compared with a set of data gathered from existing residential buildings through monitoring and surveys. Despite its shortcomings in model inputs and dataset, the study represents a considerable step toward the validation of modelling studies.

The various studies indicate that the most representative types of the building stock have been studied. However, none of these studies addressed in detail the high-rise residential buildings, a fast growing solution in large cities such as London [37]. The Tall Building Survey of 2017 [39] indicates that 30% of homes currently under construction in London are in high-rise buildings and that the number of tall buildings construction sites has increased by 68% since 2016, a solution to accommodate London's 10 million residents by 2030 [40].

Three main characteristics of high-rise residential buildings reinforce overheating. First, they are usually associated with lightweight construction methods with low thermal mass [16]. Second, a typical apartment in a high-rise residential building is expected to face one single direction, often preventing cross-ventilation. Third, market preferences have favoured large and convex glazing for better views and increased daylighting. From the design point of view, daylighting and overheating appear to be conflicting as stated in the UK Approved Document Part L and the Code for Sustainable Homes. Add-on, passive cooling shading devices, green roofs or shutters are considered an additional and unnecessary expenditures by developers [16].

Finally, in terms of available overheating design assessments, the Zero Carbon Home has published a detailed review of all tools and methodologies to assess overheating risk [41]. For building regulation compliance in the UK, the government's Standard Assessment Procedure (SAP), provides Appendix P that contains a procedure to verify the tolerable level of solar gains. The SAP tool considers the impact of geographical location, external temperature, thermal properties of the building tested and solar gains. However, the tool has a limited capacity to deal with complex interactions between the factors that contribute to overheating. The other category of tools used for overheating assessment is Dynamic Simulation Modelling (DSM). In comparison with the SAP's, the DSM tools can include a wider range of design parameters and can predict a large number of parameters (Internal temperatures, energy consumption, HVAC systems etc.). It is also possible to apply different overheating standards such as the CIBSE benchmarks or the adaptive thermal thresholds. They are considered as more sophisticated and requires a certain level of training.

The reviewed literature states that residential buildings are at a higher risk of overheating and indicates that top floor apartments are most vulnerable to thermal discomfort risks than lower ones. The relationship between the floor position of the apartments in the building and the overheating risk is however not clearly defined. In response, first, this study aims to clarify this relationship by assessing the overheating risk in a high-rise residential building in relation to floor position, orientation, glazing ratio and thermal mass. Second, this study investigates the relationship between daylighting and overheating risk to determine a balanced design solution. Finally, since the current overheating assessment tools in the UK have either limited abilities to predict the overheating risk at an early design stage, or are too complex to use, this study aims to develop simple design tools that rely on a rapid assessment of overheating risk and daylight performance at an early design stage in high-rise residential buildings.

2. Method

2.1. Standards and Benchmarks

Several methodologies and benchmarking thresholds are used to quantify and assess the overheating in buildings. While it is recognized that assessing overheating in existing buildings is more problematic due to monitoring limitations, overheating benchmarking at design stage is easier. In the UK, the Chartered Institution of Building Services Engineers (CIBSE) has developed weather data [42] and static overheating criteria [43] for this purpose. The criteria have evolved over time and the current CIBSE static criteria classify a building that is overheating if the internal temperature exceed 28 °C and 26 °C, in living rooms and bedrooms, respectively, for more than 1% of the occupied hours. Using the CIBSE benchmark, the UK industry assesses overheating at design stage through thermal dynamic simulation software, such as The Integrated Environmental Solution (IES) Virtual Environment (VE).

The other way of assessing the overheating is to follow the American Society of Heating, Refrigerating and Air-Conditioning Engineers (ASHRAE) method [44,45]. The adaptive thermal comfort gives a temperature threshold that changes with the mean of the ambient temperature and the sensitivity of the occupants. Hence, the adaptive thermal comfort might seem more suitable to assess overheating in free running buildings. Thermal comfort is achieved via an adaptive model that relies on the interaction between the occupant and its local climate, season and weather. The downside of the adaptive thermal model is that it was based on the collection of field data from office buildings, which might make it not suitable for residential buildings.

Both static and adaptive models are arguably associated with some limitations. The CIBSE Guide TM52 (2013) explained that "all comfort standards have problems, because they try to give precise definitions when the phenomenon they are describing is inherently imprecise" [46]. Consequently, the data from the simulations are reviewed by both the CIBSE benchmarks and the adaptive method. For the bedrooms, a lower comfort temperature associated with sleeping will be used, 26 °C for the CIBSE threshold and the Cat I for the adaptive methodology. For the living room, a higher comfort temperature will be used to assess overheating, 28 °C and Cat II for both CIBSE and EN15251.

From the introduction, it is understood that daylighting design is an additional barrier when considering overheating free design. For this reason, the daylighting represents the second focus to gauge the issue and draw appropriate conclusion in parallel with the thermal comfort analysis. Looking at the current practise, the building regulation of the UK has no specific requirement for daylighting in dwellings. However, the Approved Document Part L [47] advise designers to follow closely the guidance given by the BS 8206-2 Code of practise for daylighting [48] to maintain a good level of daylighting. The BS 8206-2 set minimum average daylight factor for different spaces in dwellings. Kitchens, living rooms and bedrooms should achieve a minimum of 2%, 1.5% and 1% for the average daylight factor, respectively. Therefore, a special consideration should be given to the size of windows and glazed area to provide adequate level of daylighting while controlling solar gains to avoid overheating.

2.2. Simulation Software and Weather Files

Currently, a long list of energy and thermal dynamic simulation software exist in the market. Among the most developed and used in academia as well as industry are Energy Plus, ESP-r (Energy Simulation Software tool), and IES VE (Integrated Environmental Solutions). For this investigation, IES VE is chosen for its ability to model radiative, conductive and convective heat exchange between the external and internal environment and construction elements, and its capacity to dynamically simulate the occupancy, solar and air densities, heating elements, cooling systems and air flows. It is also one of the most used software for energy and thermal modelling in the UK industry and the academia studies [35,49].

The dynamic simulations is carried out using the "control" weather file published by the PROMETHEUS [32] project for the current climatic conditions in London Islington. The London Islington location has been selected as it represents an urban area with a low green space density and a high Land Surface Temperature (LST) [9]. In addition, for the climate change impact investigation, each simulation will be repeated with a projected Test Reference Year (TRY) for 2030, 2050 and 2080 with high emission scenario (a1fi) at 90th percentile probability. Even if the Design Summer Year (DSY) weather files have been designed especially for the overheating studies, TRY weather files represent more appropriate data for this investigation [50].

2.3. Base Model Characteristics

The study is based on a dynamic simulation of a real building in the City of London, representing a typical high-rise residential building (Central circulation core, high glazing ratio on the facades, low thermal mass, high occupancy) and incorporating best practise, which is exceeding building regulations for new residential buildings. The model is validated using the same building characteristics as the real building (Floor layout, concrete and steel structure, internal finished, mechanical ventilation, density of occupants). However, the windows and the external walls have been slightly improved to meet higher energy saving targets (see Section 2.4). Figure 1 represents the floor layout of the modelled building. The simulated apartments are modelled between two similar floors. The total floor area of the apartments varies between 60 and 70 m^2. The orientation, the glazing ratio and the shading from the adjacent buildings are considered as variables. Table 1 and Figures 2 and 3 illustrate the variation of solar radiation at different floor levels and orientation.

Figure 1. Floor layout of the base model.

Table 1. Solar Radiation (S.R) at 0%, 25%, 50%, and 75% Adjacent Shading Height (A.S.H).

Orientation	S.R at 0% Shaded (kWh/m²·a) Flat Position: Top Floor	S.R at 25% Shaded (kWh/m²·a) Flat Position: Middle Top Floor	S.R at 50% Shaded (kWh/m²·a) Flat Position: Middle Bottom Floor	S.R at 75% Shaded (kWh/m²·a) Flat Position: Bottom Floor
North	360	320	190	130
East	590	455	260	160
South	750	560	360	185
West	560	440	270	150
A.S.H (m)	0	12	17	33

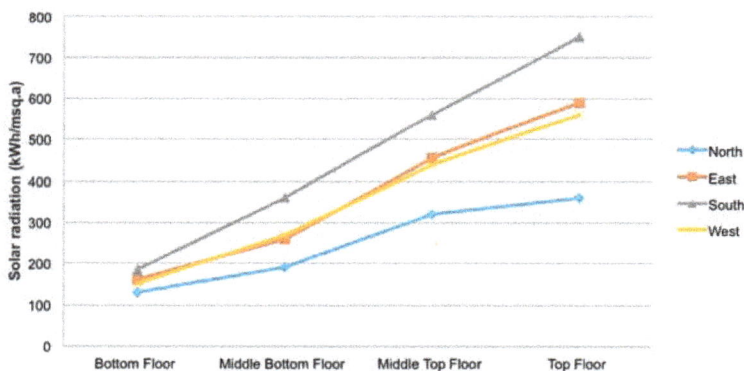

Figure 2. Solar radiation in each orientation (North, East, South, and West) and at each flat position (Bottom, Middle Bottom, Middle Top, and Top).

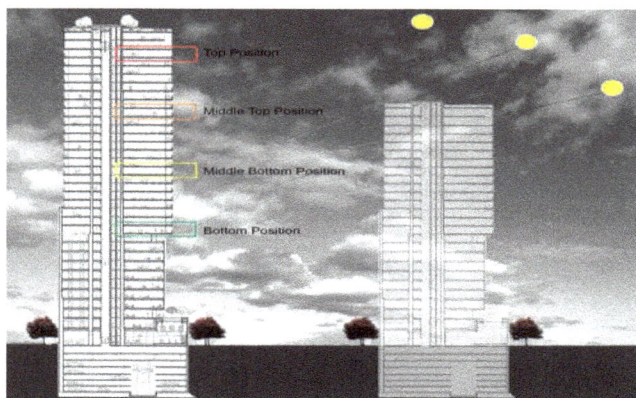

Figure 3. Apartment positions illustration.

2.4. Design Combinations and Model Characteristics

The simulation study is based on a single reference building and a set of combination of design parameters. The high-rise residential building contains single- and double-sided apartments at different orientations. All apartments were assumed to have a high level of insulation modelled with both mechanical and natural ventilation strategy. The floor location of the apartments in the building is the main variable. To support the study, variation in the orientation, the thermal mass and the glazing ratio are taken into consideration in the actual and future weather scenarios.

Table 2 summarises the design variables that are the base of 1536 cross-combinations. It should be also noted that, in a real context, the design process is more complex and might be different from the assumptions taken in this study and might consider other variables and alternative design parameters.

Table 2. Design combinations of the investigation/Design parameters of the study.

Design Variables						
Typology	Floor Position	Orientation	Thermal Mass	Glazing Ratio	Ventilation	Weather
Single-sided	Bottom Middle bottom	North East	Low	40% 60%	Mechanical	Control TRY 2030
Double-sided	Middle top Top	South West	Medium	80%	Natural	TRY 2050 TRY 2080

The building fabric and its thermal properties are primordial to the overheating investigation in dwellings. There is a considerable need to reduce the energy consumption for the space heating demand. Consequently, a super insulation strategy is used for the study. The thermal performance of the models follows PassivHaus standards, which represents an improvement of around 40% from the 2010 England and Wales Building Regulations. The thermal properties used for the models are resumed in Table 3. In addition, a lightweight (60 kJ/m^2·K) and a medium weight (140 kJ/m^2·K) thermal mass are tested. The glazing ratio is considered as another variable to gauge the thermal and daylighting performances. Glazing ratios of 40%, 60% and 80% are tested for the different models to assess the impact of the glazing on the overheating risk. For the daylighting simulation, 70% reflectance for the walls is considered, which represents an approximation for a typical light colour reflectance.

Table 3. Thermal properties of construction elements.

Construction Elements	U-Values (W/m²·K)	g-Value	Glazing Lighting Transmittance	Window Frame Factor (%)	Window Proportion Length/Height
External walls	0.15				
Roof	0.1				
Floor	0.1				
Windows	0.85	0.6	0.7	30	0.33

The CIBSE Guide A [43] is used as a base to model the internal heat gains in the apartments. Three groups of internal heat gains, lighting, appliances and occupants (young couple with one child or an adult flatmate) are modelled (Table 4). The occupancy chosen will allow the overheating assessment of both single and double bedrooms.

Table 4. Modelled internal gains and occupancy profile.

Space	Internal Gain Category	Sensible Gain	Latent Gain	Occupancy Profile and Number of Occupant
Master Bedroom	People	50.2 W/person	23.6 W/person	10:00 p.m.–7:00 a.m. every day 2 people
	Lighting	18 W		6:00 a.m.–7:00 a.m. 9:00 p.m.–11:00 p.m. every day
	Appliances	20 W		6:00 a.m.–7:00 a.m. 10:00 p.m.–11:00 p.m. every day + 10% heat gains for background standby use 24 h/day
Bedroom	People	50.2 W/person	23.6 W/person	10:00 p.m.–7:00 a.m. every day 1 people
	Lighting	18 W		6:00 a.m.–7:00 a.m. 9:00 p.m.–11:00 p.m. every day
	Appliances	20 W		6:00 a.m.–7:00 a.m. 10:00 p.m.–11:00 p.m. every day + 10% heat gains for background standby use 24 h/day
Living room	People	75 W/person	55 W/person	6:00 a.m.–9:00 a.m. 5:00 p.m.–11:00 p.m. Every day; 2 people
	Lighting	36 W		8:00 p.m.–11:00 p.m. every day
	Appliances	120 W		6:00 a.m.–9:00 a.m. 5:00 p.m.–11:00 p.m. every day + 10% heat gains for background standby use 24 h/day
Kitchen	People	75 W/person	55 W/person	7:00 p.m.–8:00 p.m.
	Cooking appliances	1000 W		7:00 p.m.–9:00 p.m.
	Fridge/Freezer	31 W		24 h/day

For this study, the space heating system is modelled during the cold months (October–April) with a set-point temperature of 20 °C, which is the comfort temperature recommended by the Passivhaus Institute. The heating system is turned off during summer months to avoid any interference with the overheating assessment.

Mechanical ventilation is simulated first for more sensible comparisons. The dynamic simulation accounts for two types of air transfer: the mechanical air supply and the uncontrolled infiltration. The infiltration is modelled as a fixed flow rate (0.25 air changes per hour), which is best practise for a passivhaus building. Based on the concept of a balanced dwelling, the mechanical ventilation extracts and supplies air at an equal flow rate. The extract is from the wet rooms (bathrooms and kitchens) and the supply from the dry rooms (living rooms and bedrooms). In addition, the heat recovery (HR) system is modelled with a summer by-pass system.

As regulated by the Approved Document F [51], the mechanical ventilation is modelled with two ventilation rates. The first one is the background, which is maintained at a constant flow rate in relation to the occupancy of the apartments. Second, the boost feature, which is modelled to permit purge ventilation at late afternoon if there is an increase in heat gains. It is also considered in bedrooms for night purging when the internal temperature exceeds the comfort temperature and when the external temperature is below the internal one but not exceeding 10 °C difference. This will allow energy conservation during cold nights.

Natural ventilation is considered in a second set of simulations to mainly assess the difference between the potential ventilation of both typologies (single- and double-sided) and to compare the effectiveness of both natural ventilation and mechanical ventilation strategies to mitigate the overheating risk.

The bulk airflow is simulated based on the differences in pressure across the operable windows and the equivalent orifice area. The operable windows are modelled with an openable area of 20% and a maximum angle of 10°. Following the MacroFlo Calculation methods [52], it corresponds to 0.34 discharge coefficient. Consequently, the equivalent orifice area is given by the same documents as almost 11% of gross area. In addition, as the position of the flats represents the main variable for the investigation of this study, the exposure type of the openings will change depending on the position of the apartments in the building. It adjusts the wind pressure coefficient in relation to the degree of sheltering of the surrounding buildings.

It is very important to simulate the opening pattern of the operable windows in accordance with the occupancy profiles of the apartments. In brief, the windows will operate in the early morning, late afternoon and evening for the living room, and for the bedrooms the openings are used mainly for night purging. The windows of the bedrooms and the living rooms are designed to open at an internal temperature of 26 °C and 28 °C respectively only if the external temperature is lower than the internal one. The internal doors will be modelled to remain open.

3. Results and Discussion

3.1. Overheating and Daylighting Performance in Relation to Apartment's Floor Position

Given the floor positions of the considered apartments, it is notable that both the daylight factor and the number of hours above the overheating benchmark increase from the bottom to the top floors. As an example, Figure 4a illustrates the overheating assessment and the daylight factor analysis for a south living room at different floor positions and variable glazing ratio with low thermal mass. For top-level apartments, the daylight factors for the 40%, 60% and 80% glazing ratios exceed the British Standards' recommendations. However, the hours above the benchmark are significantly higher than the CIBSE threshold. For the bottom position, it is the opposite. While the living rooms are meeting the thermal comfort benchmark, the daylight factors are at their lowest values. In addition, Figure 4a reflects the conflict between daylighting and overheating trends. Looking at different orientations several similarities can be observed. They all illustrate the same direct relationship between the floor position, the glazing ratio and the overheating risk (Figure 4a–d).

One of the main observations is that higher flat positions and glazing ratio result in a higher overheating risk, which is in line with previous studies [31–34]. Consequently, an increase in both parameters could represent a significant threat to the internal thermal comfort. In addition, for some apartments positions, it is very difficult to meet the recommended daylighting level. As can be seen in Figure 4, the bottom positions have very low daylight factors, which do not meet the British Standards' recommendations. Even if it is meeting the overheating requirement, the low daylighting factors in the bottom positions represent one of the consequences of the vertical urbanization. In terms of design optimization, finding the right solution that meets both thermal comfort and daylight is challenging.

However, this multidisciplinary study aims to optimize the design process at an early design stage. This is of particular relevance to practitioners and researchers who need simplified, quick-to-run overheating and daylighting assessment tools.

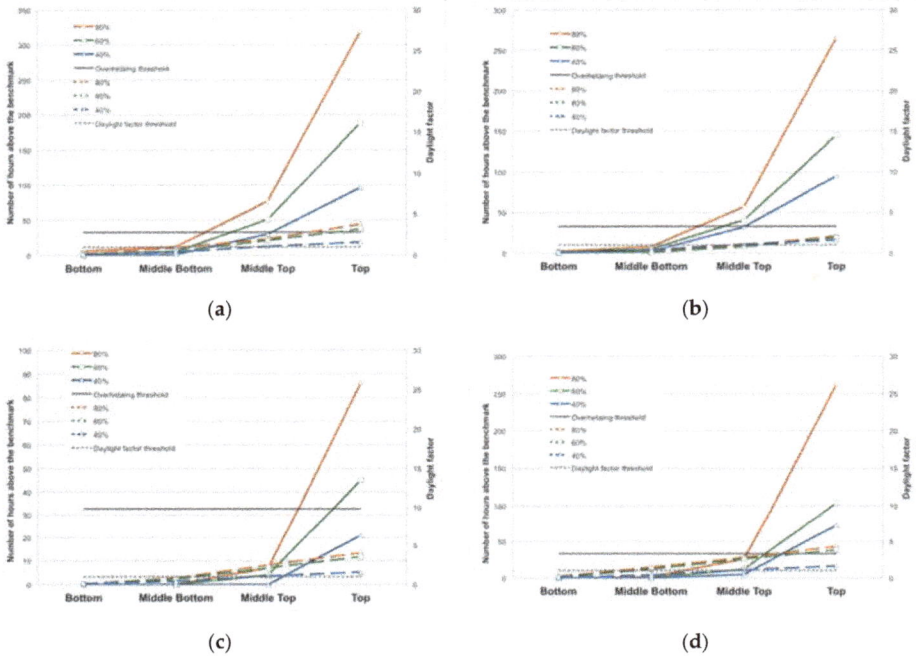

Figure 4. Number of hours above the benchmark and the daylight factor for the mechanically ventilated living rooms at 40%, 60% and 80% glazing ratio and at bottom, middle bottom, middle top and top position (London, Islington, and control): (**a**) south facing; (**b**) west facing; (**c**) north facing; and (**d**) east facing.

To better understand the floor position parameter and its impact on the overheating risk in high-rise residential buildings, Figure 5a shows the temperature variation of three days (21–23 July) where the flats are exceeding the overheating benchmark and the external temperature is peaking at almost 30 °C. A significant temperature difference can be observed between the same living rooms at different apartment positions. The highest living rooms are at a higher internal temperature. To explain this difference in temperature, Figure 5b illustrates the solar gains difference between the living rooms at different apartment floor position. The solar gains are much higher at top level than at bottom level because of the surrounding buildings that provide adjacent shading at lower level. That causes an increase in the total solar gains, which build up heat internally and worsen the thermal comfort. As a mitigation strategy, which is incorporated in the model, the ventilation is more frequently solicited to exhaust the heat and cool down the internal temperatures (Figure 5a).

The glazing ratio represents the second element that significantly impacts the hours above the overheating benchmark and the daylight factor. On the one hand, increasing the glazing ratio raises the overheating risk by allowing a higher conduction and solar gains (Figure 12). Looking at the ventilation at 80% glazing ratio (Figure 13), the flow rate of the mechanical system indicates that the boost feature is turned on more often and with a longer time-lapse. There is here a strong relationship between the glazing ratio, the overheating risk and the need for mitigation strategy such as the ventilation, which is in accordance with existing literature [18,53]. On the other hand, there is a direct relationship between daylighting and glazing ratio. Figures 5–9 demonstrate that the daylighting factor increases with the glazing ratio. While it seems beneficial for daylighting design to increase the glazing ratio, it can be seen as a significant threat to thermal comfort by increasing the overheating risk. As they are linked together, there is an obvious need to design for both visual and thermal comfort in the same time, and finding the optimal design solution represent the main contribution of this study.

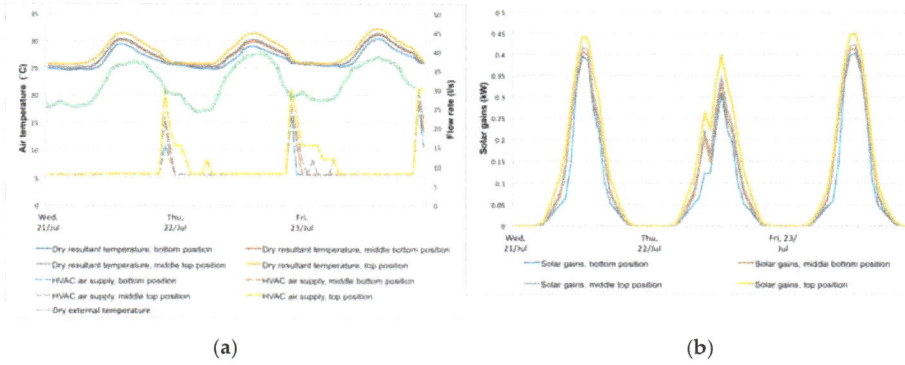

(a)   (b)

Figure 5. The influence of different flat positions in: (**a**) the air temperature, mechanical air flow rate and (**b**) solar gains of south facing living rooms at 60% glazing ratio (London, Islington, and control).

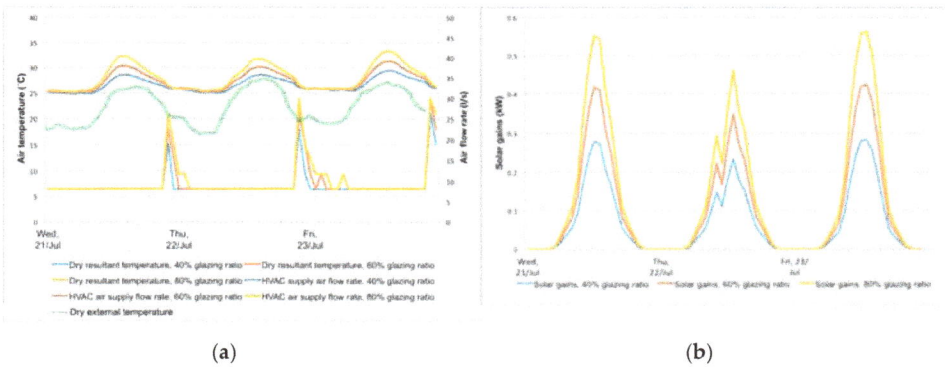

(a)   (b)

Figure 6. The influence of different glazing ratios in: (**a**) the air temperature, the mechanical air flow rate; and (**b**) the solar gains of south facing living rooms at middle top position (London, Islington, and control).

3.2. Impact of the Orientation on the Overheating and Daylighting Performances

In urban dense high-rise residential buildings, most of the apartments are expected to face one single direction. Regardless of the weather and environmental conditions, the internal environment of an apartment may react differently depending on its orientation. The results illustrate a significant difference between the number of hours above the CIBSE benchmark for north, east, south and west facing living room (Figure 7a). The north facing living rooms are at lower overheating risk than the other orientations and the South facing living rooms represent the most thermally uncomfortable spaces. In addition, the daylight factor results illustrated in Figure 7b shows that the North and East facing living rooms have a higher daylight factor than any other orientation and that the west facing living rooms have the lowest daylight factor.

An indication that daylight factor and overheating risk may be at cross-purposes has been discussed in the previous chapter. However, the orientation plays an important role in the cross-purpose relationship. For example, it can be seen that for the north orientation it is easier to meet the daylight factor while having a low number of hours above the benchmark. Conversely, for the west facing bedrooms, it is much more difficult to meet the British Standards' recommendations while avoiding overheating.

This difference in overheating risk at different orientation can be explained by looking at the solar gains (Figure 8), which are at a different intensity and occur at a different time during the day. It can be observed that the east and west facing living rooms have a peak of solar gains at almost 0.6 kW in the morning and evening respectively. While the north and south oriented bedrooms benefits from solar gains at midday with a less peaky curves. For a better understanding, the internal air temperature and people heat gains (occupancy pattern) have been included in Figure 8. It is observed that the peak in solar gains and internal air temperature for the living rooms oriented north occurs in the middle of the non-occupied period. Consequently, the living rooms overheat in the unoccupied period and lose this build up heat at late afternoon, which is significantly improving the thermal comfort for the occupant during the evening. For the south facing living rooms, the peak in solar gains is also occurring during the non-occupied period. However, the solar gains are more intense and the internal temperature increases rapidly during the beginning of the afternoon, which makes it more difficult for the ventilation to mitigate the overheating risk. Regarding the west and east orientation, the peak of solar gains is closer to the late afternoon and morning occupied hours respectively. The difference between both orientations is that the west facing living room suffers more from the heat stored during all the day, which impacts on the internal temperature during the late occupied hour. The evidence shown here explains the cause of a high number above the overheating benchmark of the south and west facing living rooms. The build-up heat during the day due to the solar and internal gains increases the internal temperature and worsens the thermal comfort, which increases the overheating risk. The orientation of the apartments can impact drastically the visual and thermal performances and finding the optimal design solution may prove to be difficult.

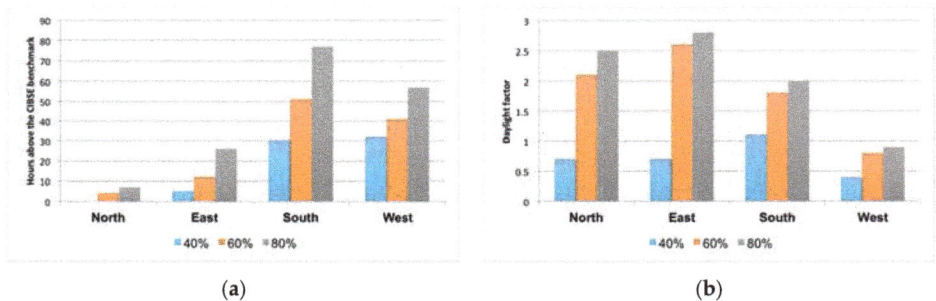

Figure 7. The influence of the orientation in: (**a**) the hours above the CIBSE benchmark; and (**b**) the daylight factor for middle top position living rooms at 40%, 60% and 80% glazing (London, Islington, and control).

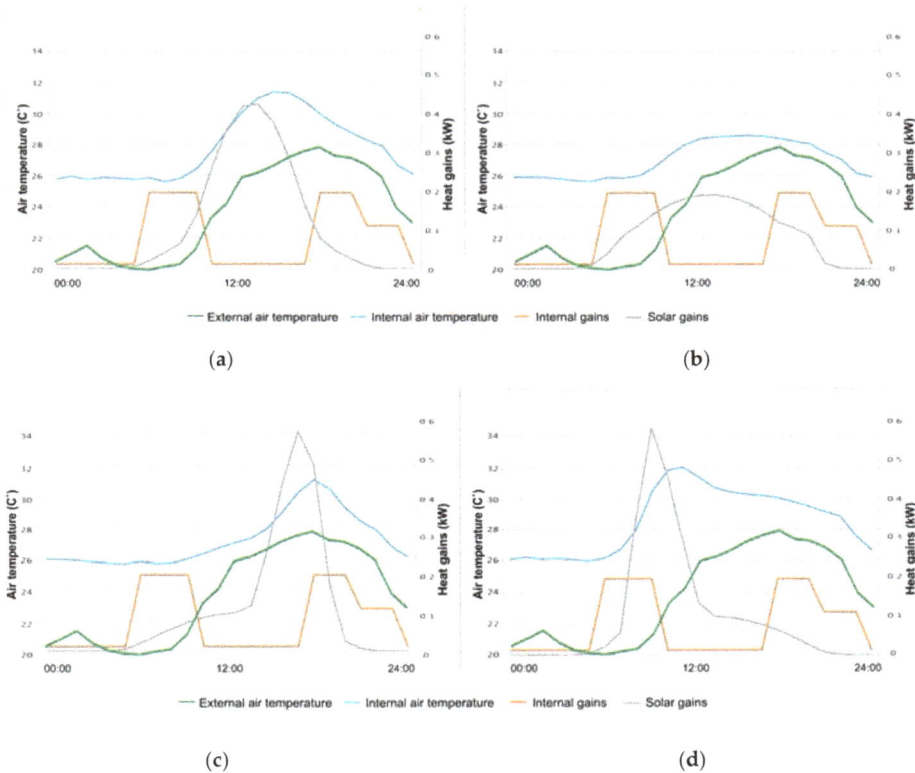

Figure 8. Internal air temperature, and solar and internal gains for the middle top living rooms at 60% glazing on 22 July (London, Islington, and control): (**a**) south facing; (**b**) north facing; (**c**) west facing; and (**d**) east facing.

3.3. Overheating, Daylighting and Ventilation Strategy

The previous simulation results were based on a mechanical ventilation strategy described in the method section, and the same models are simulated with a natural ventilation strategy. The analyses have shown that the mechanical ventilation can, in some cases, mitigate the overheating risk in the flats in an efficient way (Figures 5a and 6a). However, the mechanical ventilation strategy based on the Part F of the UK building regulation has some difficulties to purge the build up heat in higher apartment positions (top position in Figure 5a). Figure 9a illustrates the number of hours above the CIBSE benchmark for the south facing living rooms with different glazing ratios. All of the top position and most of the middle top position living rooms exceed the overheating threshold. In comparison with the living rooms, Figure 9b demonstrates that the south facing bedrooms are not at overheating risk and that the number of hours above the overheating benchmark is considerably lower. This is due to a lower exposure to internal heat gains, which can play a significant role in the effectiveness of the ventilation system. As an example, the kitchens are designed to be completely open on the living rooms. Even if both spaces are modelled independently in terms of function, they are thermally connected. The opening between the two spaces allows a complete transfer of heat and air. Figure 10 illustrates heat gains from the kitchen and the internal temperature of the living room in the south-facing apartment during 22 July, which is the middle of the overheating period (21–23 July). The first observation is that the kitchen solar gains participate in the sharp increase of the living room internal temperature during the day. Secondly, during the late afternoon, when the boost ventilation is on (Figure 11), the high cooking heat gains (Figure 10) are participating to the increase in internal temperature and the reduction of the purge ventilation's efficiency.

The earlier results, which are in agreement with previous studies [12,54], have demonstrated that there is a direct relationship between the glazing ratio, solar gains and the overheating risk. Higher glazing ratio results in a higher overheating risk. Looking at the mechanical ventilation pattern during three days in July where the south top position living rooms overheat (Figure 11a), it is noticeable throughout the increase in flow rate, that the purge ventilation feature is used during the late afternoon. It is also observed that the purge ventilation is used at a different intensity and different time duration. To mitigate the risk of overheating, the mechanical ventilation is triggered at a higher flow rate and for a longer time-lapse, which increases the energy consumption of the mechanical ventilation system. Therefore, for a reduction of both overheating risk and energy consumption, a lower glazing ratio might be considered. However, when considering daylighting, a reduction in glazing ratio seems to reduce the daylight factor, which in return might worsen the daylighting level in the apartments.

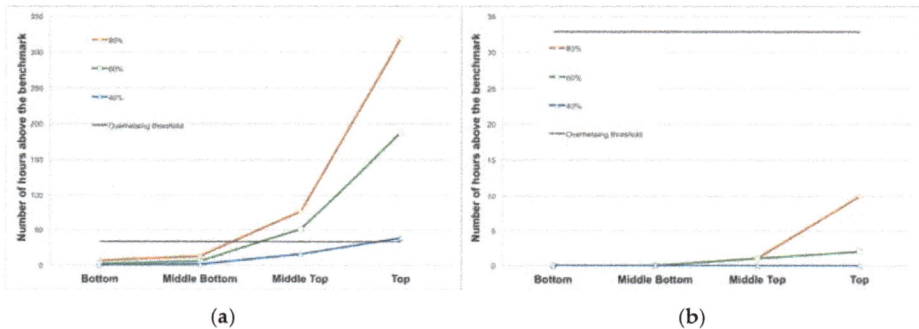

(a) (b)

Figure 9. Number of hours above the CIBSE benchmark for the mechanically ventilated south facing: (**a**) living rooms; and (**b**) bedrooms at 40%, 60% and 80% glazing ratio (London, Islington, and control).

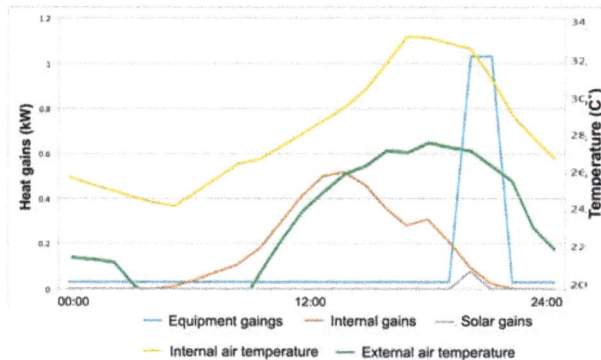

Figure 10. Internal heat gains (equipment, solar and people) and internal air temperature of the top position south facing living rooms at 80% glazing ratio on 22 July (London, Islington, and control).

Considering some issues that might occur in dense cities, such as a reduction in airflow or urban air pollution and noise level, natural ventilation strategies are being thwarted. However, when it is feasible, natural ventilation could be a very effective solution to drop the internal temperature and mitigate passively the overheating risk (Figure 11b). In most cases, the adopted natural ventilation strategy successfully mitigates the overheating risk. When comparing with the mechanical ventilation living rooms, the naturally ventilated living rooms demonstrate lower overheating exposure. Opening the windows allow a high flow rate entering the apartments, which helps reducing the overheating risk by purging the build up heat quickly.

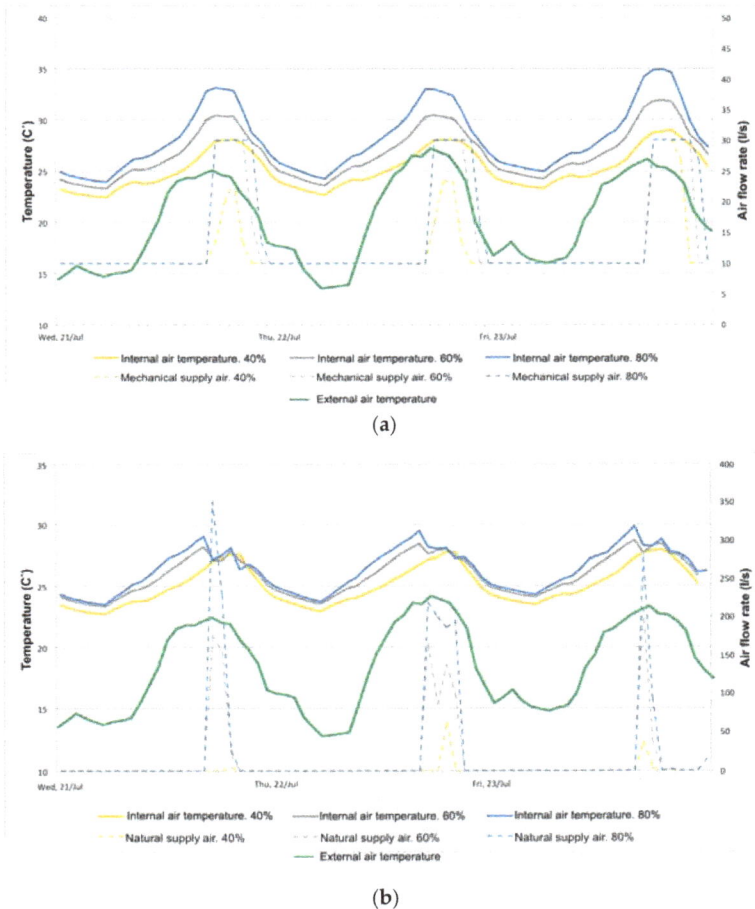

(a)

(b)

Figure 11. The influence of: (**a**) the mechanical air flow rate; and (**b**) the natural air flow rate in the internal temperature of the south facing living rooms at top position (London, Islington, and control).

Both natural and mechanical ventilation strategies have shown their effectiveness to reduce the overheating risk. However, for the apartments that are more exposed to the solar radiation, natural ventilation seems to be more effective and helps mitigating the overheating risk. Figure 12 illustrates a comparison between the number of hours above the CIBSE benchmark for mechanically and naturally ventilated south facing living rooms with low thermal mass at 60% glazing ratio. It should be noted that 60% glazing ratio has been chosen as the middle between 40% and 80%. For the top position the hours above the benchmark for the naturally ventilated living room is nine times less then the mechanical ventilated one. A considerable reduction in the overheating risk is noticed when using natural ventilation strategy. However, as described before, some surrounding conditions and safety reasons may reduce the wind speed and prohibit the occupants to open the windows. Consequently, careful decisions must be taken when designing ventilation system.

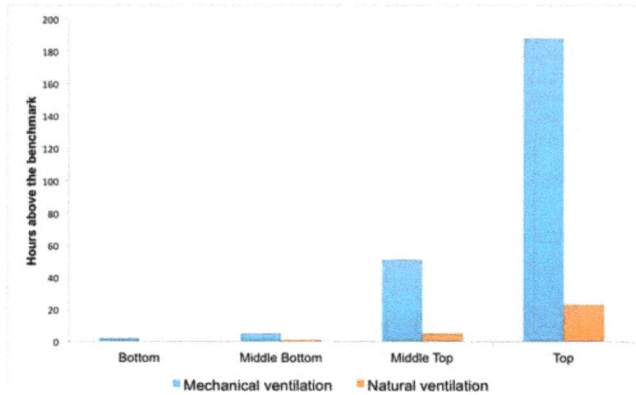

Figure 12. Number of hours above the CIBSE benchmark for both naturally and mechanically ventilated south facing living rooms at 60% glazing ratio (London, Islington, and control).

The model was designed in such a way to allow cross ventilation to occur in dual orientation apartments and hence enable airflow to cross from a space to another. This is demonstrated by the airflow rate entering the apartments and the capacity to mitigate the overheating risk. Figure 13 gives a more detailed analysis of why the dual orientation apartments perform better than the single orientation in terms of thermal comfort. Looking at one of the living rooms that are overheating the most, the west facing living rooms shows that the double-sided apartments provides a higher flow rate, which means a quicker purging capacity. It can also be seen that in late afternoon the living room in the double-sided apartments has a lower temperature. As it has been proven in previous studies, providing cross ventilation helps reducing the risk of overheating. This research highlights that a lower overheating risk give more daylighting design flexibility for dual orientation apartments, which can increase daylighting levels by increasing the glazing ratio.

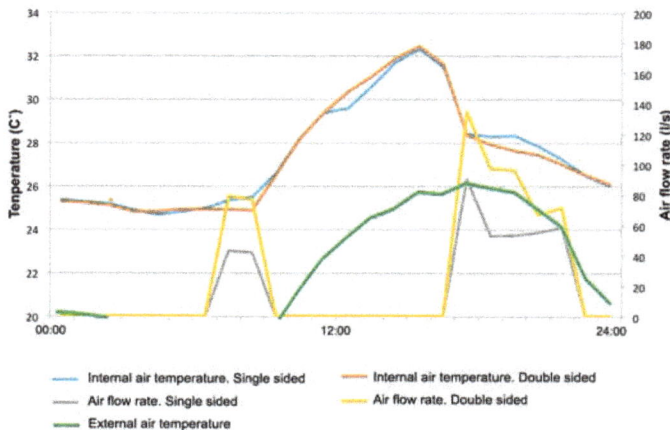

Figure 13. Internal air temperature and air flow rate for both double and single-sided apartments, west facing living rooms at 60% glazing ratio and top position on 22 July (London, Islington, and control).

3.4. Summary of Results

Based on all the results from the different design parameters' combinations, the relationship among the floor position, the overheating risk and daylighting is clear. All the temperature and daylighting data have been collected and analysed separately using a simple linear regression method in Excel to understand their relationship with the floor position. Looking at the scatter plots (Figure 14) of the results, the simple linear regression shows a coefficient of determination "R^2" of 0.7, which demonstrates a strong statistical correlation between the floor level and the predicted internal temperature. They also demonstrate that the highest apartments are more prone to overheating than the lower ones, which is in accordance with the existing literature [22,54]. The relationship between the floor level and the overheating demonstrates the need for a careful design throughout the floor levels.

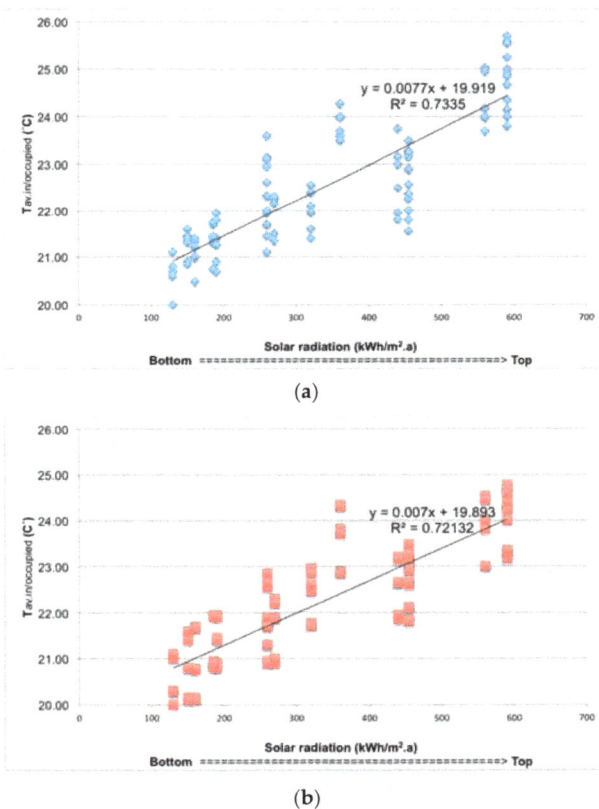

(a)

(b)

Figure 14. The average internal air temperature during the occupied hours from May to September of the: (a) mechanically ventilated apartment; and (b) naturally ventilated apartment (London, Islington, and control).

Considering just the overheating issue, it seems that having the same design parameters from the bottom to the top of a high-rise residential building can be disastrous for some flats. In the other hand, the daylighting factor has a less strong correlation with the floor level of the flats (Figure 15). Although the R^2 value of the correlation between the daylighting and the floor level is almost 0.5, the daylighting scatter plot shows a similar trends that the internal temperature figures. The daylight factor increases with the height, however, in this case, it is seen as a positive outcome. Overheating risk and daylighting are at conflict and the results shown is of a particular relevance to practitioners and researchers who aim to find the optimal thermally and visually comfortable design.

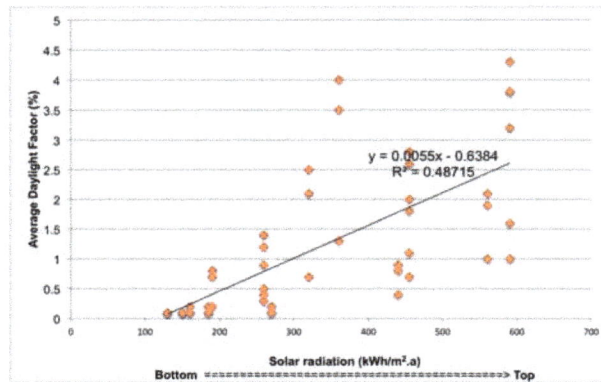

Figure 15. The Daylight Factor for both naturally and mechanically ventilated apartments (London, Islington, and control).

4. Design Assessment Tools

One of the main objectives of this study is to build simple assessment tools to help avoid overheating and meet the recommended daylighting by quickly predicting the conditions of the internal environment and rapidly assessing different design options. This section will give more details on how the tools have been created and how they can be used. The design assessment tools presented in this paper can be used only for multiple storey residential buildings in London.

The design prediction tool has been created through a multivariate linear regression method using the regression analysis tool in Excel. This method helps to build a mathematical relationship between four independents variables and one dependant variable. Three relationships are considered:

- Relationship between the design variables (orientation, floor position, glazing ratio and thermal mass) and the predicted internal temperatures of the mechanically ventilated apartments.
- Relationship between the design variables (orientation, floor position, glazing ratio and thermal mass) and the predicted internal temperatures of naturally ventilated apartments.
- Relationship between the design variables (orientation, floor position, glazing ratio and thermal mass) and the predicted daylighting factor.

Equations (1)–(3) illustrate the result of the computational process of the multivariate regression analysis. They represent a simple tool under the form of a formula that can predict the internal temperatures and the daylighting factors in relation to only four design variables. Designers and researchers can predict the internal thermal and visual conditions by inserting values for the design parameters that have the most significant impact (orientation, the floor position, the glazing ratio and the thermal mass), and the result will predict the daylighting factor and the average internal temperatures for the mechanically or naturally ventilated apartments. This can be used to quickly predict the effectiveness of a variety of design strategies and to give guidance at an early design stage.

$$T_{av.in(occup)} \approx -3.87 \times 10^{-6}\,\chi + 7.66 \times 10^{-3}\lambda + 2.55 \times \varphi - 2.63 \times 10^{-3}v + 18.65 \tag{1}$$

Equation (1). Average internal air temperature during the occupied hours of the hottest months for the mechanically-ventilated apartments.

$$T_{av.in(occup)} \approx -1.06 \times 10^{-6}\,\chi + 7.26 \times 10^{-3}\lambda + 3.26 \times \varphi - 6.80 \times 10^{-4}v + 18.51 \qquad (2)$$

Equation (2). The average internal air temperature during the occupied hours of the hottest months for the naturally-ventilated apartments.

$$DF \approx -1.92 \times 10^{-3}\chi + 6.01 \times 10^{-3}\lambda + 2.67 \times \varphi - 1.32 \qquad (3)$$

Equation (3). The Daylight Factor formula for both naturally and mechanically-ventilated apartments. where

$T_{av.in(occup)}$: Average internal air temperature during the occupied hours.

DF: Average Daylight Factor (%).

χ: Orientation, the yearly vertical solar radiation of the flat's orientation (kWh/m^2·a).

λ: Position, the yearly vertical solar radiation on the flat's facade (kWh/m^2·a).

φ: The window/wall glazing ratio of the flat (%).

v: The thermal mass of the flat (kJ/m^2·K).

In addition, a design comparison tool was generated using the result of each single alteration made on the model. Figure 16 shows whether each design combination considered in this study meets the daylight factor threshold of the British Standards and the CIBSE overheating criteria. In addition, Figure 16 compares the same design combinations using the actual and future weather scenarios, which enables the users to assess the impact of climate change on the overheating risk.

The comparison tool can be easily used at an early design stage to assess the overheating and daylighting performance of different high-rise residential buildings. The tool considers the following variables:

- Climate data: Control, TRY 2030, TRY 2050, TRY 2080
- Ventilation strategy: mechanical, natural
- Glazing Ratio: 40%, 60%, 80%
- Thermal mass: low, medium
- Orientation: north, east, south, west
- Flat position: top, middle top, middle bottom, bottom
- Type of space: living room, bedroom

Designers and Architects can easily reduce the overheating risks and increase daylighting levels in actual and future weather scenarios by assessing their early high-rise residential design models using this simple design comparison tool, which will report to them whether each apartment is:

- Meeting both overheating criteria and daylighting levels
- Meeting overheating criteria, but not daylighting levels
- Not meeting overheating criteria, but meeting daylighting levels
- Neither meeting overheating criteria nor meeting daylighting levels

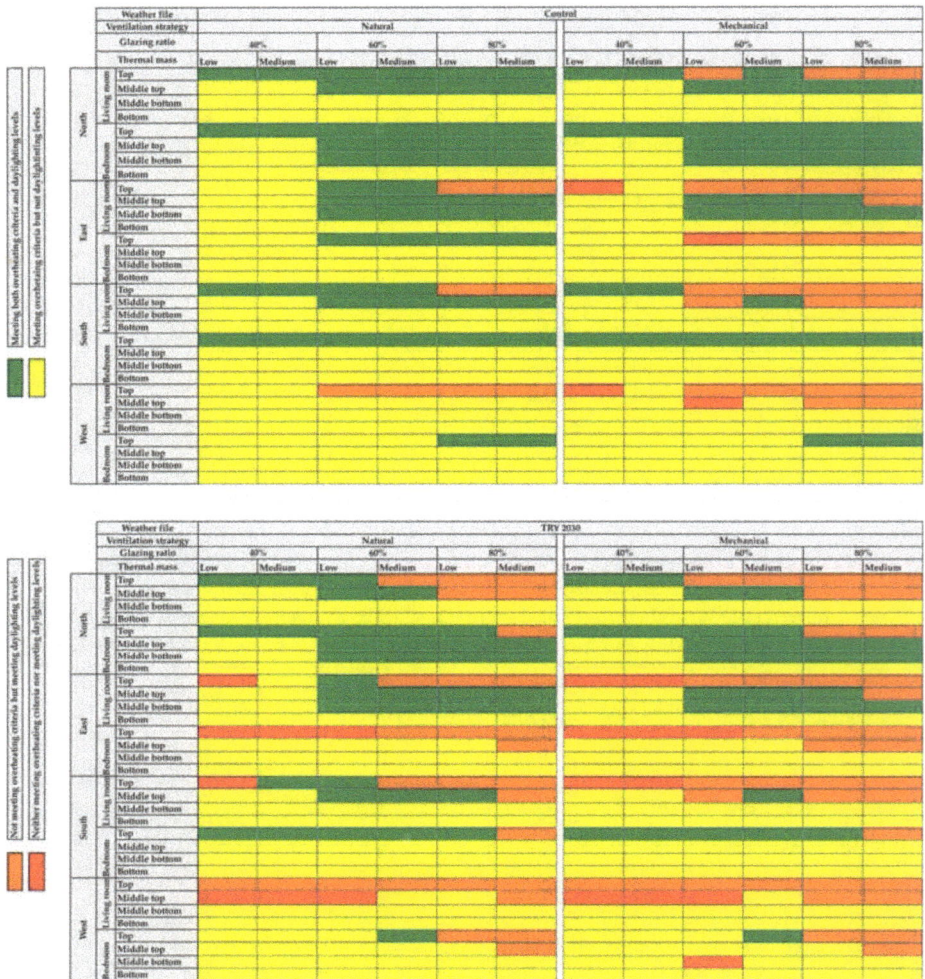

Figure 16. *Cont.*

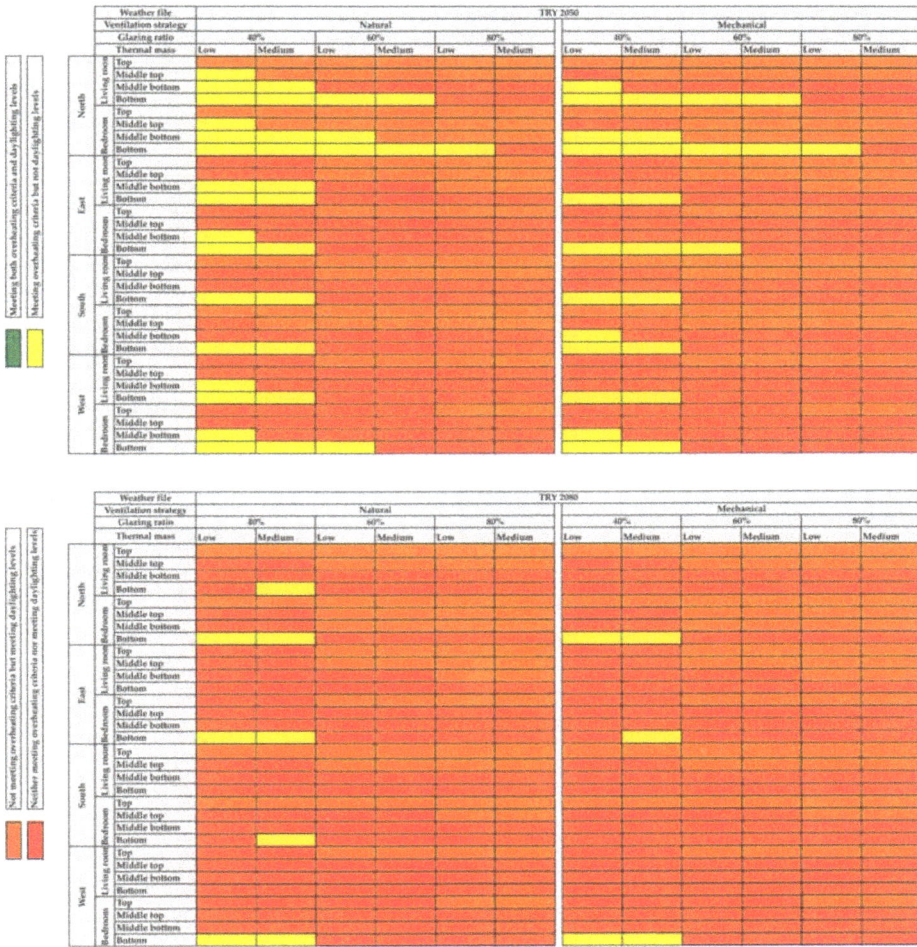

Figure 16. Overheating and daylighting comparison tool to be used at early design stage.

5. Conclusions

This paper has discussed a design optimization quest to avoid overheating and meet appropriate daylighting level in London's high-rise residential buildings. The work was based on thermal dynamic simulations to assess different design combinations while considering the typology, floor position, orientation, thermal mass, glazing ratio, ventilation strategy and the weather. The aim was to build simplified tools capable of predicting internal temperatures and daylighting levels. Furthermore, the tools give the ability to compare the overheating risk and daylight performance of different apartments in high-rise residential buildings.

The findings of the study are significant in at least two main points. First, this study has clarified that there is a strong relationship among the glazing ratio, the floor position, overheating risk and daylighting performances of apartments in London's high-rise residential buildings. When considering the same design variables, the apartments on the top floors are exposed to a higher risk of overheating and a better daylighting performance, which resulted in a major conflict between overheating and daylighting design. At lower floor positions the opposite tends to happen. The orientation adds another level of complexity. The South and West apartments were at a higher risk of overheating, while the ones on the north and east were much cooler. In terms of daylighting, the North and East apartments were preforming better than the South and West apartments. Together, the floor position, the glazing ratio and the orientation are the main parameters to consider for overheating and daylighting design optimisation.

The last key finding and the major contribution of this study is the creation of design assessment tools that help architects and researchers reduce the overheating risk and meet appropriate daylighting levels during early design stages. The prediction and comparison tools are easy and quick to run. They can provide design guidance and useful information on the thermal and visual performances of apartments in London's high residential buildings.

In terms of limitations, a validation against real monitored high-rise residential buildings can help refine the results gathered from the thermal and daylighting models. Such an exploration is beyond the scope of this study, however, further investigation could reduce the level of error and help increase the level of accuracy of the prediction and assessment tool. Another major source of uncertainty is the indoor overheating criteria. Both CIBSE fixed threshold and the adaptive thermal model are arguably associated with a certain level of limitation. On the one hand, the CIBSE fixed threshold is highly criticized for its lack of adaptation. On the other hand, the adaptive thermal comfort is purely based on data collected entirely in office buildings.

Considering the findings of this study, high-rise residential buildings are in need of different design strategies at different floor levels and orientation. The number of high-rise residential buildings is increasing significantly in London and there is an urgent need to meet the energy reduction challenges, to provide adequate indoor temperatures and to design at the recommended daylighting levels. These can be easily tackled using simple design tools at an early design stage.

Acknowledgments: The authors gratefully acknowledge the financial support in the form of fee waiver from the Special Issue Editor, Arman Hashemi and help, at early stages of this study, from Kartik Amrania, Head of Building Sustainability Department at Sweco UK.

Author Contributions: Bachir Nebia perceived the idea, carried out the simulation and co-wrote the paper. Kheira Tabet Aoul guided and participated in the writing of the paper.

Conflicts of Interest: The authors declare no conflict of interest.

References

1. Solomon, S.; Qin, D.; Manning, M.; Chen, Z.; Marquis, M.; Averyt, K.B.; Tignor, M.; Miller, H.L. *Contribution of Working Group I to the Fourth Assessment Report of the Intergovernmental Panel on Climate Change, 2007*; Cambridge University Press: Cambridge, UK, 2007; ISBN 978 0521 88009-1 Hardback.
2. Robine, J.M.; Cheung, S.L.; Le Roy, S.; Van Oyen, H.; Herrmann, F.R. *Report on Excess Mortality in Europe During Summer 2003*; European Commission, Directorate General for Health and Consumer Protection: Brussels, Belgium, 2007.
3. Johnson, H.; Kovats, R.S.; McGregor, G.; Stedman, J.; Gibbs, M.; Walton, H.; Cook, L.; Black, E. The impact of the 2003 heat wave on mortality and hospital admissions in England. *Health Stat. Q.* **2005**. [CrossRef]
4. Zero Carbon Hub. *Overheating in Homes, the Big Picture, Zero-Carbon Hub, Full Report*; Zero Carbon Hub: London, UK, 2015.
5. Public Health England. *Heatwave Plan for England, Protecting Health and Reducing Harm from Severe Heat and Heatwaves*; Public Health England and National Health Service: London, UK, 2015.
6. Gabriel, K.M.A.; Endlicher, W.R. Urban and rural mortality rates during heat waves in Berlin and Brandenburg, Germany. *Environ. Pollut.* **2011**, *159*, 2044–2050. [CrossRef] [PubMed]
7. Laaidi, K.; Zeghnoun, A.; Dousset, B.; Bretin, P.; Vandentorren, S.; Giraudet, E.; Beaudeau, P. The impact of heat islands on mortality in Paris during the August 2003 heat wave. *Environ. Health Perspect.* **2011**, *120*, 254–259. [CrossRef] [PubMed]

8. United Nations, Departement of Economics and Social Affairs (UN, DESA). *World Urbanization Prospects, 2014 Revision*; United Nations, Departement of Economics and Social Affairs: New York, NY, USA, 2015.
9. Zero Carbon Hub. *Overheating Risk Mapping, Evidence Review*; Zero Carbon Hub: London, UK, 2015.
10. Committee on Climate Change (CCC). *Meeting Carbon Budgets—2012 Progress Report to Parliament*; Committee on Climate Change: London, UK, 2012.
11. Hamilton, G.; Shipworth, I.D.; Summerfield, J.; Steadman, A.P.; Oreszczyn, T.; Lowe, R. Uptake of energy efficiency interventions in English dwellings. *Build. Res. Inf.* **2014**, *42*, 255–275. [CrossRef]
12. Porritt, S.; Cropper, P.; Shao, L.; Goodier, C. Ranking of interventions to reduce dwelling overheating during heat waves. *Energy Build.* **2012**, *5*, 16–27. [CrossRef]
13. Shrubsole, C.; Macmillan, A.; Davies, M.; May, N. 100 Unintended consequences of policies to improve the energy efficiency of the UK housing stock. *Indoor Built Environ.* **2014**, *23*, 340–352. [CrossRef]
14. National House Building Council (NHBC); Zero Carbon Hub. *Overheating in New Homes F46—A Review of the Evidence*; IHS BRE Press: Watford, UK, 2012.
15. Beizaee, A.; Lomas, K.J.; Firth, S.K. National survey of summertime temperatures and overheating risk in English homes. *Build. Environ.* **2013**, *65*, 1–17. [CrossRef]
16. Lomas, K.J.; Porritt, S.M. Overheating in buildings: Lessons from research. *Build. Res. Inf.* **2017**, 1–18. [CrossRef]
17. Baborska-Narozny, M.; Stevenson, F.; Chatterton, P. Temperature in housing: Stratification and contextual factors. *Eng. Sustain.* **2015**, *9*, 1–17. [CrossRef]
18. Hulme, J.; Beaumont, A.; Summers, C. *Energy Follow-up Survey 2011, Report 7: Thermal Comfort & Overheating*; Building Research Establishment: Watford, UK, 2013.
19. Ji, Y.; Fitton, R.; Swan, W.; Webster, P. Assessing overheating of the UK existing dwellings—A case study of replica Victorian end terrace house. *Build. Environ.* **2014**, *77*, 1–11. [CrossRef]
20. Lomas, K.J.; Kane, T. Summertime temperatures and thermal comfort in UK homes. *Build. Res. Inf.* **2013**, *41*, 259–280. [CrossRef]
21. Mavrogianni, A.; Taylor, J.; Davies, M.; Thoua, C.; Kolm-Murray, J. Urban social housing resilience to excess summer heat. *Build. Res. Inf.* **2015**, *43*, 16–33. [CrossRef]
22. Mavrogianni, A.; Pathan, A.; Oikonomou, E.; Biddulph, P.; Symonds, P.; Davies, M. Inhabitant actions and summer overheating risk in London dwellings. *Build. Res. Inf.* **2017**, *45*, 119–142. [CrossRef]
23. Vellei, M.; Ramallo-González, A.P.; Coley, D.; Lee, J.; Gabe-Thomas, E.; Lovett, T.; Natarajan, S. Overheating in vulnerable and non-vulnerable households. *Build. Res. Inf.* **2017**, *45*, 102–118. [CrossRef]
24. Vellei, M.; Ramallo-González, A.P.; Kaleli, D.; Lee, J.; Natarajan, S. Investigating the overheating risk in refurbished social housing. In Proceedings of the 9th Windsor Conference: Making Comfort Relevant, Windsor Great Park, UK, 7–10 April 2016.
25. Gul, M.S.; Jenkins, D.; Patidar, S.; Menzies, G.; Banfill, P.; Gibson, G. Communicating future overheating risks to building design practitioners: Using the low carbon futures tool. *Build. Serv. Eng. Res. Technol.* **2015**, *36*, 182–195. [CrossRef]
26. Gupta, R.; Gregg, M. Using UK climate change projections to adapt existing English homes for a warming climate. *Build. Environ.* **2012**, *55*, 20–42. [CrossRef]
27. Holmes, M.J.; Hacker, J.N. Climate change, thermal comfort and energy: Meeting the design challenges of the 21st century. *Energy Build.* **2007**, *39*, 802–814. [CrossRef]
28. Mavrogianni, A.; Wilkinson, P.; Davies, M.; Biddulph, P.; Oikonomou, E. Building characteristics as determinants of propensity to high indoor summer temperatures in London dwellings. *Build. Environ.* **2012**, *55*, 117–130. [CrossRef]
29. Taylor, J.; Davies, M.; Mavrogianni, A.; Shrubsole, C.; Hamilton, I.; Das, P.; Biddulph, P. Mapping indoor overheating and air pollution risk modification across Great Britain: A modelling study. *Build. Environ.* **2016**, *99*, 1–12. [CrossRef]
30. DeWilde, P.; Rafiq, Y.; Beck, M. Uncertainties in predicting the impact of climate change on thermal performance of domestic buildings in the UK. *Build. Serv. Eng. Res. Technol.* **2008**, *29*, 7–26. [CrossRef]
31. Architecture, Engineering, Consulting, Operations, and Maintenance (AECOM). *Investigation into Overheating in Homes: Literature Review*; Department for Communities and Local Government: London, UK, 2012.
32. Arup Research + Development, Bill Dunster Architects. *UK Housing and Climat Change: Heavy vs. Lightweight Construction*; Ove Arup & Partners Ltd.: London, UK, 2005.

33. Orme, M.; Palmer, J. *Control of Overheting in Future Housing, Design Guidance for Low Energy Strategies*; AECOM Ltd.: Hertfordshire, UK, 2003.

34. Orme, M.; Palmer, J.; Irving, S. *Control of Overheating in Well-Insulated Housing*; Faber Maunsell Ltd.: Hertfordshire, UK, 2003.

35. Coley, D.; Kershaw, T. Changes in internal temperatures within the built environment as a response to a changing climate. *Build. Environ.* **2010**, *45*, 89–93. [CrossRef]

36. Good Homes Alliance (GHA). *Preventing Overheating: Investigating and Reporting on the Scale of Overheating in ENGLAND, including Common Causes and an Overview of Remediation Techniques*; Good Homes Alliance: London, UK, 2014.

37. Departement of Communities and Local Governement (DGLG). English Housing Survey. In *Headlibe Report 2013–2014*; Departement of Communities and Local Governement: London, UK, 2015.

38. Ledrans, S.; Vandentorren, P.; Bretin, A.; Zeghnoun, L.; Mandereau-Bruno, A.; Croisier, C.; Cochet, J.; Ribéron, I.; Siberan, B.; Declercq, M. August 2003 heat wave in France: Risk factors for death of elderly people living at home. *Eur. J. Public Health* **2006**, *16*, 583–591. [CrossRef]

39. New London Architecture (NLA). *London Tall Buildings Survey 2017*; New London Architecture: London, UK, 2017.

40. New London Architecture; GL Hearn. *London Tall Buildings Survey 2015*; GL Hearn Limited: London, UK, 2015.

41. Zero Carbon Hub. *Assessing Overheating Risk—Evidence Review*; Zero Carbon Hub: London, UK, 2015.

42. Chartered Institution of Building Services Engineers (CIBSE). *Guide J: Weather, Solar and Illuminance Data*; Chartered Institution of Building Services Engineers: London, UK, 2002.

43. Chartered Institution of Building Services Engineers (CIBSE). *Guide A: Environmental Design*, 8th ed.; Chartered Institution of Building Services Engineers: London, UK, 2015.

44. American National Standards Institute and American Society of Heating Refrigeration and Air-conditioning Engineers (ANSO/ASHRAE). *Standard 55 Thermal Environmental Conditions for Human Occpancy*; American National Standards Institute and American Society of Heating Refrigeration and Air-conditioning Engineers: Atlanta, GA, USA, 2013.

45. British Standards Institution. *Indoor Environmental Input Parameters for Design and Assessment of Energy Performance of Buildings Addressing Indoor Air Quality, Thermal Environment, Lighting and Acoustics*; European Committee for Standardization: Brussels, Belgium, 2007.

46. Chartered Institution of Building Services Engineers (CIBSE). *Guide TM52: The Limits of Thermal Comfort: Avoiding Overheating in European Buildings*; Chartered Institution of Building Services Engineers: London, UK, 2013.

47. Department for Communities and Local Government (DCLG). *Approved Document L1A*; Department for Communities and Local Government: London, UK, 2010.

48. Raynham, P.C.O.C. *Lighting for Buildings—Part 2: Code of Practice for Daylighting*; British Standard BS 8206-2; British Standards Institution: London, UK, 2008.

49. Reeves, T.; Olbina, S.; Issa, R.R.A. Guidelines for Using Building Information Modeling for Energy Analysis of Buildings. *Buildings* **2015**, *5*, 1361–1388. [CrossRef]

50. Eames, M.; Kershaw, T.; Coley, D. On the creation of future probabilistic design weather years from UKCP09. *Build. Serv. Eng. Res. Technol.* **2011**, *32*, 127–142. [CrossRef]

51. Department for Communities and Local Government (DCLG). *Approved Docuement F*; Department for Communities and Local Government: London, UK, 2010.

52. Integrated Environmental Solutions Limited (IES). *MacroFlo Calculation Methods*; Integrated Environmental Solutions Limited: Glasgow, UK, 2011.

53. Coley, D.; Kershaw, T.; Eames, M. A comparison of structural and behavioural adaptations to future proofing buildings against higher temperatures. *Build. Environ.* **2012**, *55*, 159–166. [CrossRef]

54. Paul, R. *Avoiding Overhetaing in Medium & High Rise Apartments*; Silcock Dawson & Partners: Leeds, UK, 2011.

sustainability

MDPI

Article

Energy and Economic Performance of Plant-Shaded Building Façade in Hot Arid Climate

Mahmoud Haggag *, Ahmed Hassanand Ghulam Qadir

Architectural Engineering Department, College of Engineering, UAE University, PO Box 15551, Al-Ain, UAE; ahmed.hassan@uaeu.ac.ae (A.H.); ghulam.qadir@uaeu.ac.ae (G.Q.)
* Correspondence: mhaggag@uaeu.ac.ae; Tel.: +971-50-563-8461

Received: 15 September 2017; Accepted: 30 October 2017; Published: 6 November 2017

Abstract: The use of vegetated walls and intensive plantation around buildings has increased in popularity in hot and arid climates, such as those in the United Arab Emirates (UAE). This is due to its contribution towards reducing the heat gain and increasing the occupants' comfort levels in spaces. This paper examines the introduction of plant-shaded walls as passive technique to reduce heat gain in indoor spaces as a strategy to lower cooling demand in hot arid climate of Al-Ain city. Experimental work was carried out to analyze the impact of using plantation for solar control of residential building façades in extreme summer. External and internal wall surface and ambient temperatures were measured for plant-shaded and bare walls. The study concluded that shading effect of the intensive plantation can reduce peak time indoor air temperature by 12 °C and reduce the internal heat gain by 2 kWh daily in the tested space. The economic analysis reveals a payback period of 10 years considering local energy tariff excluding environmental savings.

Keywords: energy efficiency; hot climate; plant-shaded wall; thermal performance

1. Introduction

Building energy efficiency, an important design problem, is increasingly being achieved though optimal passive design approach. Through careful selection of building layout and materials in context of local climate, energy consumption can be reduced retaining thermal comfort [1]. Shading buildings by artificial or natural means is proven technique to save energy in hot climates [2]. Solar shading in subtropical regions of China saved 26.06% and 24.42% compared to Low-E windows and fabric roller shades respectively [3]. One extension of the same consideration is plant shading as the impact of plant shaded walls creates a balance between energy performance and aesthetic appearance. The amount of shading to be provided can be easily manipulated from very light 10% coverage to very dense 80% coverage by appropriate selection of height and density of the plants. Additionally, a layer of air is trapped within the vegetation thus limits the movement of heat through the wall of the building by means of evapotranspiration [4–6].

The level of thermal performance depends on plant covering percentage, density & width of plant foliage, type & size of the trees and orientation of the plant-shaded walls [7]. The plant orientation is reported to influence energy savings where the west and east directions are recommended to grow trees understandably to provide shade for most of the day time [8]. McPherson and Simpson [9] further determined the impact of orientation coupled with distance between the plants and façade. They concluded that energy savings of plant-shaded walls are positive on east and west directions up to 12.2 m, neutral on south direction due to summer savings being offset by winter losses, not affected on north due to direct radiation not being blocked and not impacted beyond 12.2 m at any façade due to shadow not reaching the building [10].

A study conducted in Amman, Jordan with trees on the west and south faces of a typical residential house reported an energy saving of up to 23.96% in hottest month of July [11]. The summers in Amman

(May to September) are hot and dry with cool evenings, the hottest month being July with average maximum temperature of 33 °C and humidity around 38% whilst the coldest weather is in December and January with an average temperature of 10 °C [12]. A study conducted in the US cities namely Minneapolis, Charlotte and Metro Orlando where the temperature ranges are 2.9 °C to 12.9 °C, 9.3 °C to 21.6 °C and 17.9 °C to 22.9 °C respectively [13] determined the effect of climatic conditions on energy savings of plant shaded wall. They concluded substantially varying energy savings in different climatic zones being 14 kWh in Minneapolis, MN, 25 kWh Charlotte, NC and 44 kWh in Metro Orlando, FL [14,15]. In Australian summer and spring conditions, it was found that, tree shade reduced wall surface temperatures by up to 9 °C and external air temperatures by up to 1 °C [6] The combined cooling and shading effect of trees is reported to save up to 50% of building air-conditioning costs [4]. The effect of vegetated living wall installed on a school building façades employing local plants was studied in Al-Ain, UAE during the peak summer (July). The study concluded that the installed plantation can reduce peak time indoor air temperature by at least 5 °C, and reduce the peak air conditioning energy demand by 20% [16].

In winter, the plant shaded wall creates a buffer against the wind, which reduces the energy loss associated with indoor heating. Therefore, using plants to shade walls have year round thermal advantages, with economic and energy-saving benefits [17]. In winter, under a Nordic Climate, the green roofs held the moisture content below the critical volume (15–20%), and can thus improve roof insulation during freezing [18]. In the Oceanic climate during the winters the green facade showed moderate reduction of heat losses, and its energy balance was found to be 20% higher than an orthodox facade [19]

Apart from energy savings, the plants enhance the environmental quality of the urban ecosystem, bringing benefits in terms of human health [20]. Moreover, tree shade helps reduce glare and transmits diffused light coming from the sky. When the sound wave hits the plants a vibration of its elements occurs converting and dissipating sound energy into heat. The plants eventually absorb sound waves and reduce noise pollution converting it into a pleasant sound of its own in the presence of breeze [21].

In local context, the UAE forests cover 741,000 acres (300,000 ha) area [22] planted with native and exotic tree and shrub species including gafas bush, desert hyacinths and common acacia vegetation [23]. Al-Ain, has a desert climate with year-round sunshine characterized by scarce rainfall and high levels for temperature. In summer (May to September), the weather is very hot, with daytime temperatures swinging between 35 °C and 50 °C. During the winter (December to February), the daytime temperatures swings between 25 °C and 35 °C. Rainfall is infrequent and falling mainly in winter, with an annual average rainfall of 10 cm [24]. The solar irradiance (yearly average global horizontal irradiance is in excess of 20 MJ/m^2/day [25], which makes air conditioning necessary to maintain acceptable indoor comfort levels. Although proven for energy savings through previous research, economic aspects of plant shaded walls have not been studied in worst case scenarios. The present study considers the worst case scenario of extremely hot ambient, rare rain falls, total mechanical irrigation and completely desalinated water without using ground water that involves additional cost although in the extremely subsidized electricity tariff rates. The present study employs local cost of planting and maintaining trees, purchasing materials, pruning, pest and disease control and irrigation [26] which is the theme of current article.

2. Experimental Set Up

As part of an experimental work, two identical semi-attached housing units have been selected to investigate the thermal performance of plant-shaded walls in the hot climate of Al-Ain City. As shown in Figure 1, the external wall facing south of the first house is unshaded (bare wall); however, the external wall of the second house is shaded with non-deciduous trees (plant-shaded wall), both walls having a south facing area of 12 m^2. Non-deciduous shade trees, or evergreens, do not drop their leaves during the year unlike deciduous trees, which lose their leaves in winter (Figure 2).

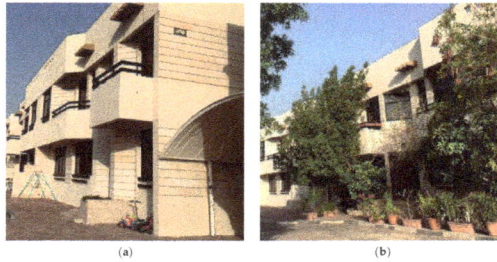

Figure 1. Plant-shaded façade in comparison with unshaded façade of identical housing units, Al-Ain, where (**a**) Unshaded building façade (bare wall), (**b**) Plant-shaded façade (plant-shaded wall).

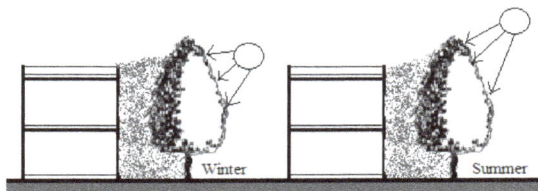

Figure 2. Non-deciduous trees provide shade in summer and winter seasons that help to conserve energy consumption in hot climate.

The tested space areas, glass ratio, furniture, construction materials, internal finishing and ventilation systems are identical in both cases. The external walls of the case studies are constructed from hollow concrete blocks with thickness of 20 cm. The internal surface of the walls is covered with white stucco, however the external surface is cladded with light color stone and stucco, using wet-fixation method (without thermal insulation layer). The glass windows nearly cover 60 percent of the building façades.

In order to truly represent the prevalent weather conditions, i.e., higher irradiance and higher heat load, and avoid the intervention from occupancy of the houses, the summer holiday season was selected to test unoccupied buildings. The experiments were conducted from the end of June to mid-August to study the impact of plantation as heat insulators in extreme hot weather and determine the resulting cooling effect indoors in residential spaces. To determine the temperature regulation effect of plant-shaded wall on indoor spaces; temperatures at four locations were recorded for both façades, using "DaqPRO" Omega data loggers: (a) outdoor ambient air temperature (1 m away from the external wall); (b) external surface temperature; (c) internal surface temperature; and (d) indoor air temperature (1 m inside from the internal wall) as shown in Figure 3. Considering spatial temperature distribution, the temperatures at each side (a–d) were measured at three points separated by a distance of 1 m each, and the average of the three is plotted.

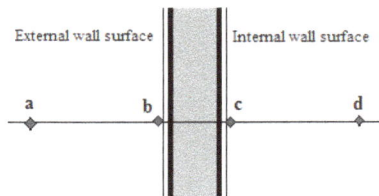

Figure 3. Positions of thermocouple wires for measuring wall surface and ambient temperatures.

The weather data of solar irradiation, ambient temperature and wind speed were consistently measured with a time steps of 10 min to achieve a uniform profile during the experiments. The accuracies of the measurement set up are provided in Table 1 below.

Table 1. Measurement ranges and accuracies of the instruments used in experiments.

Measurement Parameter	Device	Model No.	Measurement Range	Accuracy
Solar radiation	Apogee Pyranometer (1)	SP-110	-	±1%
Data acquisition	DAQ-PRO (2)	DAQ-PRO	-	±0.02%
Temperature	RS Pro Thermocouple K-type	363-0389	−75 °C to 250 °C	±1.5 (°C)
Ambient temperature	Star meter weather station	WS1041	−40 °C to 60 °C	±1%
Wind speed	Star meter weather station	WS1041	Up to 50 ms^{-1}	±1%

3. Results and Discussion

3.1. Temperature Decrement and Time Lag

To understand the impact of plants on the heat transfer through the façade into the building, the temperature evolution at the outer surface, inner surface and indoor ambience is presented for three representative days in Figures 4–6, respectively. The figures highlight three important outcomes of plants in terms of (1) temperature decrement; (2) time lag to reach peak temperature; and (3) duration of peak temperature. The bare wall (reference) and the plant shaded wall responded to the incoming radiation differently. The radiation reached the bare wall un-interrupted compared to the plant-shaded wall wherein the radiation was partially blocked by the plants.

External surface temperature of the bare wall raised to peak in quick fashion where it stayed a few hours longer compared to the plant-shaded wall during day time for the three presented days in July as shown in Figure 4. It shows that, at peak time, the external surface temperature reached around 56 ± 0.5 °C on most of the first day, while the temperature on the external plant-shaded wall peaked around 51 ± 0.5 °C for the same day showing a decrement of around 5 °C. The external surface of plant-shaded wall reached a peak at 3:30 p.m., while the bare wall reached the same temperature at 11:30 a.m., thus the plant-shaded wall showed a time lag of 4 h. Similarly, the plant-shaded wall remained at 50 °C for 30 min, while the bare wall remained at or above the same temperature for at least 8 h. All the three factors described represent the cooling effect produced by the plant shaded wall. At nighttime, the bare wall showed an increased rate of cooling yielding a lower temperature compared to the plant shaded wall attributed to the heat retention of the plant-shaded wall. A similar behavior can be observed for rest of the two days presented.

The reduced external surface temperature on the shaded wall naturally resulted in a reduction of the internal wall surface temperature compared to the internal wall surface temperature of bare wall shown in Figure 5. The internal surface temperature of the bare wall peaked at 50 ± 0.5 °C, while that of the plant-shaded wall peaked at 47 ± 0.5 °C for the same day, showing a drop of 3 °C. Similarly, the temperature increased on the plant-shaded wall with a time delay, when compared to the bare wall. The bare wall again remained at or above the maximum temperature achieved by the plant-shaded wall for at least 6 h from 2 p.m. to 8 p.m. A similar behavior can be observed for the rest of the two days presented.

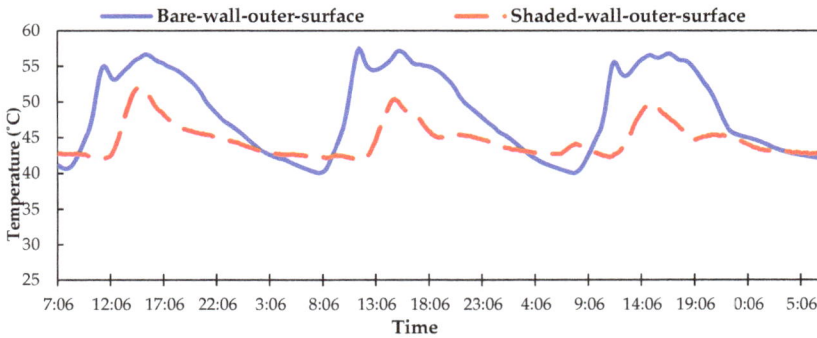

Figure 4. Temperature evolution at the front surface of the plant shaded and the bare wall.

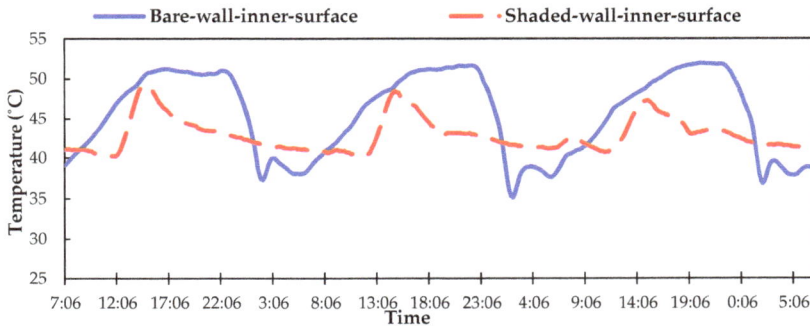

Figure 5. Temperature evolution at the interior surface of the plant-shaded and the bare wall.

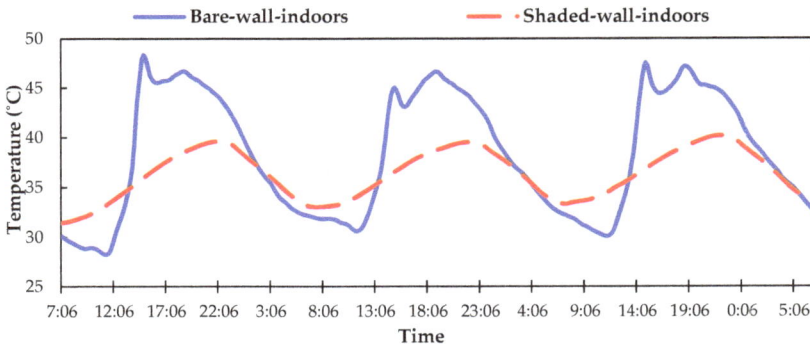

Figure 6. Temperature evolution of indoor air for plant-shaded and bare wall.

As shown in Figure 6, the indoor air temperature in the case of plant-shaded walls always showed a decrement, time lag and reduced duration of peak temperature compared to that of bare wall confirming the cooling effect produced by the plants. The peak indoor air temperature in the case of bare wall reached 47 °C, while that of the plant-shaded wall reached 39 °C, showing a drop of 8 °C. A very important finding is that the indoor temperature reached its peak very late in the evening, at 8 p.m., compared to that of bare wall at 2 p.m., showing a time lag of 6 h. A similar behavior can be observed for the rest of the two days presented. At nighttime, indoor air in the bare wall showed an increased rate of cooling, yielding a lower temperature compared to the plant shaded wall attributed to the heat retention of the plant shaded wall.

Figure 7 shows the temperature drop achieved at any time at the front surface and the indoors by the plant-shaded wall. The peak temperature drop was observed at about 11:00 a.m. at the outer surface with the values in the range of 12–14 °C, which is slightly higher than the maximum temperature drop of 9 °C reported in the weather condition of Australia [11] primarily due to summer in UAE being stronger than that in Australia, contributing to better shading performance. The peak indoors temperature dropped by 10–12 °C which is higher than the temperature drop of 5 °C achieved by green wall in the same climatic [16]. The reason can be that the trees acts as a wider buffer between radiation in facades and retard heat through natural heat dissipation into ambience, while in case of green wall, the façade is in intimate contact with the plantation that reduces heat dissipation in the ambient. The temperature drop of 3.5 °C on outer wall and 2 °C on indoor ambient was achieved, which is slightly higher than the ambient temperature drop of 1 °C reported in Australian summer, due to difference of intensity of the summer [11]. The temperature drop achieved by the plant-shaded wall over a long period of time is shown in the Appendix A, which reveals that the plant-shaded wall consistently achieved a similar temperature drop over an extended time in summer.

Despite the fact that the indoor ambient temperature of the plant-shaded façade is lower, the indoor air temperature still did not reach the comfortable temperature of 26–28 °C. This means that, in such a hot climate, the use of plant-shaded walls alone will not be enough, and mechanical cooling systems would be necessary to produce a comfortable indoor climate. The main benefit of using plant-shaded walls, however, comes from the energy savings and the reduced peak air conditioning demand.

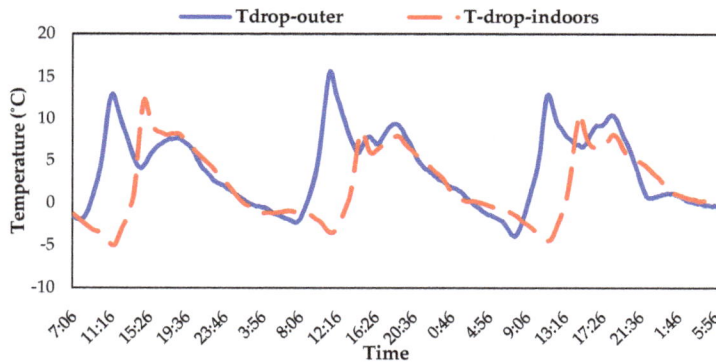

Figure 7. Temperature drop achieved by plant-shaded wall compared to the bare wall.

3.2. Heating Prevention

The temperature drop achieved by the shaded wall at the interior surface eventually transmitted less heat to indoors compared to the bare wall. The heat transmitted indoors (Q) is calculated by applying combined convection and radiation heat loss (U) represented by Equation (1). The U value is represented by $U = 5.9 + 3.4\,v$ in the air velocity (v) at 0.5 m·s^{-1} [27].

$$Q_i = hA \sum_{i=1}^{i=n} (T_{surf,i} - T_{air,i}) \tag{1}$$

where A is the surface area of the internal wall contributing to heat release (12 m^2), T_{surf} is the internal surface temperature, T_{air} is the internal air temperature and i is time step being $i = 1, 2, 3, \ldots, n$.

The transient heat transmitted at any time step is presented in the Figure 8. Predominantly, the bare wall transmits heat from the wall to indoors (heat gain-positive value), while it only transmits from indoors to wall (heat loss-negative value) late at night for a short time. In case of the plant-shaded wall, the heat transmission is always in the positive direction representing heat gain by the building. It also highlights that the plant cover acts as a heat barrier to leave outdoors at night time, which can be exploited in the moderate winter season to eliminate the meagre heating load in Al-Ain. The heat transmitted by each interior wall is monotonic with the cooling load of the building, considering that both tested spaces apply the same cooling and ventilation system.

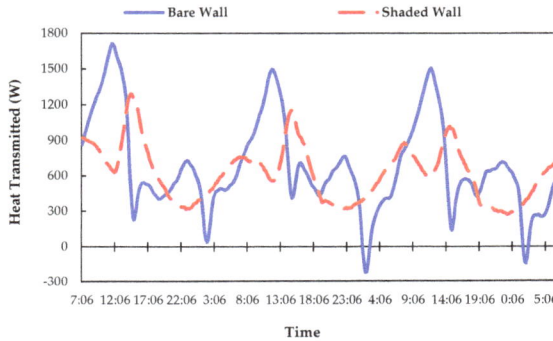

Figure 8. Heat tranmitted indoors by interior wall surface by bare wall and the plant shaded wall.

The difference of heat transmitted by bare wall and plant-shaded wall can be considered as cooling load capacity savings. The peak cooling capacity required by the bare wall reached 1.7 kW, which was reduced to 1.3 kW in the case of the plant-shaded wall. Since both the spaces have the same type of air conditioning system installed, net heat gain in certain duration of time can be regarded as cooling demand of the building. The net cooling demand is calculated for three different time frames, i.e., morning till peak time (2 p.m.), day time only (7 p.m.), and on 24 hourly basis, as shown in Figure 9. The plant-shaded wall always demanded less cooling compared to the bare wall, thus saving cooling load. Cooling load saved by the plant-shaded wall till the peak time was 3.5 kWh, which was reduced to 2.9 kWh on the day time basis and 1.5 kWh on the 24 h basis. Since the cooling demand season prevails from March to October in Al-Ain, cooling demand savings per year can be extrapolated by multiplying the daily saving by 245 cooling days, resulting in 367 kWh/year, which is much higher than the reported values of 44 kWh in Metro Orlando, FL [14,15] predominantly due to hot summer condition sin UAE.

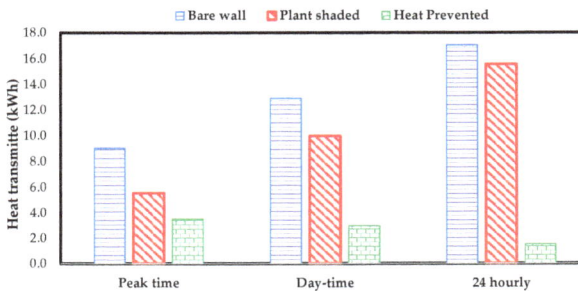

Figure 9. Net cooling demand over certain time frames for bare wall and plant-shaded wall indoor spaces.

4. Cost Benefit Analysis

Several studies have proved the economic incentives of these green techniques. However, the main question that should be asked here is "are vegetated and plant-shaded greenery systems economically sustainable in the local context?" To answer this question, the study investigates a cost-benefit Analysis of the plant-shaded facade of the case study, considering personal economic benefits as well as the life span of the building skin. As described by Perini and Rosasco, personal economic benefits are directly related to energy saving for air conditioning, improvement of real estate value, and durability of building façades [28]. At this stage, the environmental benefits of air quality improvement, carbon reduction, climate and biodiversity improvements, habitat creation, sound control and urban heat island are not included.

Initial and installation costs, maintenance and running costs of the plant-shaded building façade are considered in the study and compared mainly with cooling load reduction and property value addition. Installation costs of the plant-shaded façade were obtained from product firms and companies available in the UAE. The initial and installation costs cover plants and growing media, irrigation system and water for irrigation. As shown in Table 2, the initial cost of the analyzed greenery system was 20 US\$/m^2, while the installation costs reached 11 US\$/m^2. Irrigation using a PVC pipe network and automatic control system costs 4.6 US\$/m^2 and the local cost of used water was 3.4 US\$/m^2.

Maintenance and running costs depend mainly on the type of greenery system. Plant-shaded façade requires low maintenance in comparison with direct and indirect vertical greening systems. For plant-shaded façade, maintenance covers mainly plant pruning and can be carried out every year. For indirect greening systems combined with planter boxes, maintenance also covers water pipe replacement and plant species substitution. For living wall systems, plant replacement, removal, and transport to landfill was added to the maintenance costs [28,29]. Façade cladding renovation can be added to the maintenance cost [28]; however, it is not included in this analysis due to its variation and life cycle.

By reducing the surface temperature of a building façade and using appropriate insulation techniques, such as water proof wall panels, vegetated and plant-shaded wall techniques can protect building surfaces and extend the lifespan of the building skin [16]. This protection comes mainly from keeping rain off the building while allow moisture to escape, reducing the expansion and contraction of building materials and protecting walls against wind and solar radiation, which might affect building materials. The use of plantation on and around the building façade reduces the frequency of the maintenance service of building skin, depending on the quality of wall surface cladding, and the environmental condition. It was estimated that without vegetation, the renovation frequency of the building façade varies between 25 and 35 years. However, the use of plantation lengthens the coating lifetime of 15 years [28].

The economic performance of the plant-shaded façade is generally calculated by the mean of three indicators [28]: the Net Present Value (the discounted value of the total costs and benefits that occur within the period of life considered); the Internal Rate of Return (the annual percentage rate of return on investment; and Payback Period (the number of years from which the total revenue equals or exceeds the total costs). The present study employs payback period through energy savings and increased rental value included as benefit while ignoring the environmental benefits and inflation rates at this stage.

A living space of the housing, combined with an average façade area of 12 m^2 was tested and the total cost was calculated as shown below and summarized in Table 2.

The benefit of plant-shaded wall is also calculated through the energy savings achieved by reduced cooling load and increased rental rate.

Cooling load reduction = 370 kWh/year

Average yearly rent for the tested space with the extra external green area = 3500 US\$

Increased rental rate = 4%

1/4th of the façade is plant shaded so the increment applies to 25% of the rent.
Added property value of tested space due to greening = 0.04 × 3500 × 0.25 = 35 US$
Local cost of electricity = 320 AED/MWh = 87.12 US$/MWh
Energy cost savings = 87.12 × 0.37 = 32.23 US$/year
Total savings = saving on rental + energy cost saving = 35 + 32.23 = 67.23 US$/year
Payback period = 696/67.23 = approx. 10 years

Table 2. Cost and benefits of the plant-shaded wall per tested space.

Category	Cost in US$ (1 US$ = 3.67 AED)	Category	Benefit
Initial cost of plants and growing media	240 US$	Energy savings	32.23 US$/year
Installation cost	132 US$	Added property value	35 US$/year
Initial cost of irrigation system	55.2 US$	Air quality improvement	Ignored
Water cost for irrigation	40.8 US$/year	Carbon reduction	Ignored
Maintenance cost	108 US$/year	Street noise reduction	Ignored
Plantation space use cost	120 US$/year	Urban heat island	Ignored
Total cost	58 US$/m^2	Total Benefit	67.23
Tested façade area per floor = 12 m^2			
Total cost for the tested façade per floor (12 × 58)	696 US$	Payback Period	10 years

As shown in Table 2, the plant-shaded wall system is economically sustainable with a payback period of 10 years. The payback period employs unsubsidized local electricity rate. The most favorable economic conditions take place when the payback period is low. In comparison with other greenery techniques, where the payback periods are usually higher than 10 years, e.g., 14–20 years for indirect green façade combined with planter boxes and steel mesh [28,30]. The authors expect to achieve a reduced payback period once the social and environmental benefits are included as subject of future study to make the use of greening systems financially viable.

5. Conclusions

The use of vegetation on and around building façades has gained increasing popularity in many cities for improving thermal performance in buildings and reducing negative environmental impacts. Plant-shaded wall technique was tested in Al-Ain, the Garden City, to increase energy efficiency and reduce cooling load in residential buildings. The study finds that plant-shaded façades can reduce the yearly cooling energy by up to 0.37 MWh compared to the unshaded façades. An indoor temperature reduction of 12 °C at the peak time and 2 °C on average on a 24-h basis was achieved in the month of July. The decreased temperature was achieved by the shading and insulation effect of low-conductivity plant foliage that rendered less thermal energy gain eventually reducing cooling demand of the building. The economic analysis reveals that the payback period of plant-shaded greenery system was 10 years, in the case of unsubsidized energy tariffs. A further reduction of the payback period is expected once the environmental savings are included, which is a subject for future study.

Acknowledgments: The authors would like to express their appreciations to the College of Engineering at the UAE University for funding this research project.

Author Contributions: Mahmoud Haggag envisioned the idea; Ahmed Hassan and Mahmoud Haggag conducted the experiments; Ahmed Hassan analyzed the results; and Ghulam Qadir collected the literature survey. The paper was written as a joint effort.

Conflicts of Interest: The Authors confirm no conflict of interest.

Sustainability **2017**, *9*, 2026

Appendix A

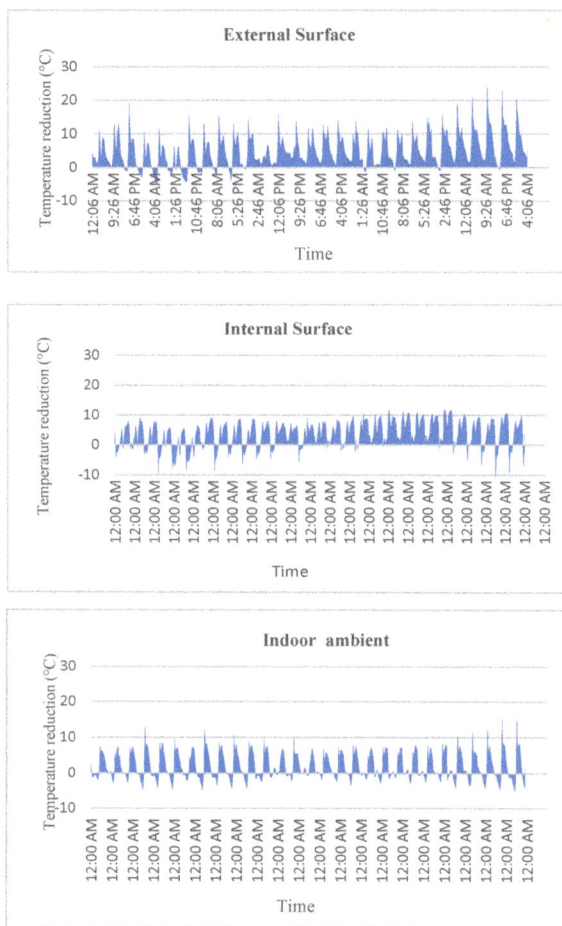

Figure A1. Temperature reductions (°C) in external surface, internal surface & ambient temperature.

References

1. Pacheco, R.; Ordónez, J.; Martínez, G. Energy efficient design of building: A review. *Renew. Sustain. Energy Rev.* **2012**, *16*, 3559–3573. [CrossRef]
2. Yao, J.; Zheng, R.Y. Determining a practically optimal overhang depth for south facing windows in hot summer and cold winter zone. *Open House Int.* **2017**, *42*, 89–95.
3. Yao, J. Effect of a novel internal roller shading system on energy performance. *J. Green Build.* **2014**, *9*, 125–145. [CrossRef]
4. Sassi, P. *Strategies for Sustainable Architecture*; Francis & Taylor: New York, NY, USA, 2006; ISBN 10 0-203-48010-4 (ebk).
5. Shahidan, M.F.; Shariff, M.K.; Jones, P.; Salleh, E.; Abdullah, A.M. A comparison of *Mesua ferrea* L. and *Hura crepitans* L. for shade creation and radiation. *Landsc. Urban Plan.* **2010**, *97*, 168–181. [CrossRef]

6. Berry, R.; Livesley, S.J.; Ayeb, L. Tree canopy shade impacts on solar irradiance received by building. *Build. Environ.* **2013**, *69*, 91–100. [CrossRef]

7. Eumorfopoulou, E.A.; Kontoleon, K. Experimental approach to the contribution of plant-covered walls to the thermal behaviour of building envelopes. *Build. Environ.* **2009**, *44*, 1024–1038. [CrossRef]

8. Hes, D.; Dawkins, A.; Jensen, C.; Aye, L. A Modelling Method to Assess the Effect of Tree Shading for Building Performance Simuation. In Proceedings of the International Building Performance Simulation Association, Sydney, Australia, 14–16 November 2016.

9. Mc Pherson, E.G.; Simpson, J.R. Potential energy savings in buildings by an urban tree planting programme in California. *Urban For. Urban Green.* **2003**, *2*, 73–86. [CrossRef]

10. Akbari, H. Shade trees reduce building energy use and CO_2 emissions from power plants. *Environ. Pollut.* **2002**, *116*, 119–126. [CrossRef]

11. Abdul Aziz, D.M. Effects of Tree Shading on Building's Energy Consumption. *J. Archit. Eng. Technol.* **2014**, *3*, 135. [CrossRef]

12. Weather and Climate. Average Monthly Weather in Amman. Available online: https://weather-and-climate. com/average-monthly-Rainfall-Temperature-Sunshine,Amman,Jordan (accessed on 18 October 2017).

13. U.S Climate Data. Available online: https://www.usclimatedata.com/ (accessed on 17 October 2017).

14. Hwang, W.H.; Wiseman, P.E.; Thomas, V.A. Tree planting configuration influences shade on residential structures in four U.S. cities. *Arboric. Urban For.* **2015**, *41*, 208–222.

15. Hwang, W.H.; Wiseman, P.E.; Thomas, V.A. Simulation of shade tree effects on residential energy consumption in four U.S. cities. *Cities Environ.* **2016**, *9*, 2.

16. Haggag, M.; Hassan, A.; Elmasry, S.K. Experimental study on reduced heat gain through green façades in a high heat load climate. *Energy Build.* **2014**, *82*, 668–674. [CrossRef]

17. Hopkins, G.; Goodwin, C. *Living Architecture: Green Roofs and Walls*; Csiro Publishing: Clayton, Australia, 2011; ISBN 9780643103078.

18. Collins, S.; Kuoppamaki, K.; Kotze, D.J.; Lü, X. Thermal behavior of green roofs under Nordic winter conditions. *Build. Environ.* **2017**, *122*, 206–214. [CrossRef]

19. Djedjiga, R.; Belarbib, R.; Bozonne, E. Experimental study of green walls impacts on buildings in summer and winter under an oceanic climate. *Energy Build.* **2017**, *150*, 403–411. [CrossRef]

20. Pandit, R.; Laband, D.N. Energy savings from tree shade. *Ecol. Econ.* **2010**, *69*, 1324–1329. [CrossRef]

21. Horoshenkov, K.V.; Khan, A.; Benkreira, H. Acoustic properties of low growing plants. *J. Acoust. Soc. Am.* **2013**, *133*, 2554–2565. [CrossRef] [PubMed]

22. El-Keblawy, A.; Al Rawai, A. Impacts of the invasive exotic Prosopis juliflora (Sw.) D.C. on the native flora and soils of the UAE. *Plant Ecol.* **2017**, *190*, 23–35.

23. Sinimar. Biota in UAE Fauna and Flora. 2017. Available online: http://www.sinimar.eu/en/biota-in-uae-fauna-and-flora/#axzz4vYn29Bob (accessed on 10 October 2017).

24. Taleb, M. Natural ventilation as energy efficient solution for achieving low-energy houses in Dubai. *Energy Build.* **2015**, *99*, 284–291. [CrossRef]

25. Mokri, A.; Aal Ali, M.; Emziane, M. Solar energy in the United Arab Emirates. *Renew. Sustain. Energy Rev.* **2013**, *28*, 340–375. [CrossRef]

26. United States Environmental Protection Agency, Using Trees and Vegetation to Reduce Heat Islands. 2017. Available online: https://www.epa.gov/heat-islands/using-trees-and-vegetation-reduce-heat-islands#4 (accessed on 15 August 2017).

27. Tiwari, A.; Sodha, M.S. Performance evaluation of solar PV/T system: An experimental validation. *Sol. Energy* **2006**, *80*, 751–759. [CrossRef]

28. Perini, K.; Rosasco, P. Cost-benefit analysis for green façades and living wall systems. *Build. Environ.* **2013**, *70*, 110–121. [CrossRef]

29. Ottelé, M.; Perini, K.; Fraaij, A.L.A.; Haas, E.M.; Raiteri, R. Comparative life cycle analysis for green façades and living wall systems. *Energy Build.* **2011**, *43*, 3419–3429. [CrossRef]

30. Dunnett, N.; Kingsbury, N. *Planting Green Roofs and Living Walls*; Timber Press: Portland, OR, USA, 2008.

sustainability

MDPI

Article

The impact of Lighting on Vandalism in Hot Climates: The Case of the Abu Shagara Vandalised Corridor in Sharjah, United Arab Emirates

Emad Mushtaha *, Ranime Ayssar Nahlé, Maitha Bin Saifan and Hasim Altan

Department of Architectural Engineering, College of Engineering, University of Sharjah, P.O. Box 27272, Sharjah, UAE; u00032147@sharjah.ac.ae (R.A.N.); architecture@sharjah.ac.ae (M.B.S.); hasimaltan@gmail.com (H.A.)
* Correspondence: emushtaha@sharjah.ac.ae

Received: 17 September 2017; Accepted: 30 October 2017; Published: 7 November 2017

Abstract: This study mainly discusses how the immature behaviour of a part of the society, resulting in vandalism, affects the building aesthetics and design features in the districts of the city of Sharjah, in the United Arab Emirates (UAE). Initially explaining the term "vandalism" in itself, this study goes on to debate on the reasons behind vandalism, its different types, and its effects on the environment. Throughout the discussion, studies of the relationship between vandalism and reflectivity are examined, considering how the characteristics and features of the buildings affect vandalism. Three methodology tools were used: a questionnaire, an Integrated Environmental Solution Virtual Environment (IESVE) software program, and illuminance measurements. Simulation scenarios of the current situation of Abu Shagara were performed, which took into account several options with respect to wall material, flooring material, and types of lighting. All in all, ten simulation cases were conducted and compared, which allowed the identification of the best simulation scenario. The type of lighting had a greater impact on the simulation scenario results than the type of wall and flooring materials. The type of lighting varied as per its polar grid and light distribution.

Keywords: vandalism; lighting; outdoor design; illuminance; IESVE simulation; UAE

1. Introduction

Vandalism is generally defined as the wilful or malicious destruction, injury, disfigurement, or defacement of property [1]. Graffiti is included as a type of vandalism in a community, since it degrades the social status of the community and diminishes the value of the properties [2]. A study conducted by Mushtaha et al. [3] recognized three main factors for the enhancement of vandalism in different housing areas, namely, the housing ecology, the social features of the residential systems, and the management system. Different studies have focused on diverse determinants regarding the structure and design of the spaces in the housing societies, particularly the perspective of visibility via dwellings' windows and the accessibility by the habitants. The perspective of visibility and the perception of accessibility were also evaluated to enhance and strengthen the perception of property and ownership in youngsters involved in vandalism.

People engage in the crime of vandalism when they purposely damage the property of others without their permission. Even though, to some citizen's view, graffiti is an artistic expression that can be seen through nature, it is still considered an illegal act that is punishable by law. Graffiti art, billboard emancipation, and crop circles are a few of the common examples of vandalism. Criminal vandalism comes in different forms. Many gang cultures use graffiti in the public properties of inner cities as a means of marking their territories. In the United States of America, the local governments have taken drastic legal measures to prevent it. However, according to research, such legal act has proven to be ineffective [4]. The products of vandalism are usually seen on building structures, street signs, billboards, tunnels, bus stops, cemeteries and other public spaces [5]. Vandalism that is left unpunished encourages people to persevere with it. In criminological research, it has been proved that vandalism is caused by different motives. According to the sociologist Stanley Cohen [6], there are six different types of vandalism, including the acquisitive vandalism, the peer pressure, and the vandalism motivated by the coolness of disobeying authority. Other types of vandalism are the tactical vandalism, the ideological vandalism, the vindictive vandalism, the pay vandalism, and the malicious vandalism [6].

The relationship between lighting and vandalism has not yet been highlighted by any research, though the topics of lighting and safety have been mentioned. Peña-García et al. [7] indicated that the main objective of public lighting is to ensure the safety of people, property, and goods. The impact of public lighting on pedestrians' well-being and perceived safety was studied by focusing on two basic aspects: (i) the average illuminance on the ground; (ii) the colour of the lights. The goals were achieved by conducting a survey in the street while the lighting was working, as well as by using illuminance measurements. Lighting design guidelines were introduced at the end of the study for lighting designers, urban planners, and especially city administrators as a way to enhance the safety and well-being of the citizens.

Abd-Razak et al. [8] investigated safety and lighting within two Malaysian universities' campuses by conducting a questionnaire without lux measurements. Based on the feedbacks received from the students, the study found that there were several unsecured locations in the campuses, like roads, walkways, and parking areas. These locations were considered as the most risky areas because they received less illumination compared to other locations in the campuses.

It seems clear that increased street lighting reduces the fear of vulnerability to criminal action in city residents, regardless of age or gender [9]. Peña-García et al. [7] indicated that this is an extremely complex problem that underscores the controversy regarding illuminance levels and safety.

In urban planning, lighting has become a topic of active debate with regard to safety, the vitality of social relations, and energy consumption. Urban lighting is defined as "the totality of all lighting in a city's public realm", and it includes street lighting as well as light from advertising, building interiors, or other artificial sources. A successful public realm allows interactions between people, enhances the economic vitality, and stimulates investments [10].

The function and role of urban lighting are based on the feelings of safety that a space provides at nighttime [11]. This study introduced new ways to improve community street lighting using the approach of Crime Prevention through Environmental Design (CPTED) in three Korean communities. It proposed design alternatives for community street lighting to enhance natural surveillance and the feelings of safety. To find ways for enhancing community safety at night, the study considered lighting standards and CPTED guidelines, using Relux Pro for a simulation analysis.

The authors of the present study believe that light not only enhances safety in a city, but also reduces vandalism because it allows a strong surveillance. As there has been no specific study so far to investigate the impact of lighting levels on vandalism, the present study intends to overcome this deficiency.

In order to enhance the lighting levels in a vandalised corridor and to analyse the resulting effects on vandalism, several space simulations corresponding to 10 alternatives, with different materials and lighting fittings, were performed in this study. This study was conducted in a vandalised corridor in the city of Sharjah (Figure 1). First, the study covers a theoretical discussion that links vandalism and lighting; next, in Section 2, it describes the study area, the written questionnaire that was used (Appendix A), the illuminance measurements, the measuring tape, and the IESVE as a simulation tool; later, in Section 3, titled "Investigated Area and Analysis", it suggests several lighting alternatives to improve the lighting levels in the vandalised corridor, and presents the results of the simulations. Finally, the "Simulation Results" section analyses each case separately and compares it with the baseline of the study and with previously described cases.

1.1. Local Standards for Street Lighting

National codes were considered as a reference and guidance for lighting alternatives, using the IESVE software package, 2017 version, for simulation. National codes stated that urban centres and public amenity areas are used by pedestrians, cyclists, and drivers. In such places, the lighting of the road surface for traffic movement is neither the main consideration, nor the only consideration, because the functions of lighting in urban centres and public amenity areas are concerned with the improvement of public safety and security, as well as with the attractiveness of the night-time environment. To fulfil these functions, a master plan should be produced to realize some or all of the objectives, which include: providing safety for pedestrians between the moving vehicles, deterring anti-social behaviour, ensuring a safe movement of vehicles and cyclists, matching the lighting design and equipment to the architecture and environment, controlling illuminated advertisements and integrated floodlighting (both permanent and temporary), illuminating the roads and the directional signs, blending light from private and public sources, limiting light pollution, and protecting lighting installations from vandalism. This list of objectives and the individual nature of each site ensure that there is no standard method of lighting urban centres and public amenity areas, nor any universally applicable recommendations. What can be given are some general recommendations for the illuminance to use in cities and town centres, although even these may need to be adjusted for a particular site, depending on the ambient environment, street parking, etc. The lighting recommended for crosswalks, pedestrians, or cyclists is greater than 15 lux [12]. In accordance with lighting installation regulations [13], and Peña-García et al. [7] stated that the average illuminances for the five streets examined in their studies should all be above 15 lux. Therefore, in this study, the authors will provide alternatives based on these references.

1.2. Significance of the Study

All elements of a lighting installation contribute to the architecture or the exterior design of a space, area, street, and/or facility. When deciding what sort of lighting to employ, the understanding of the use of space is important. The dimensions, finishes, textures, and colours of the luminaires, lit and unlit, should be considered if the desired atmosphere is to be achieved [12]. Farrington et al. [14] stated that improved street lighting could reduce crime by up to 20% and, thus, it is an important factor in safety. Blöbaum et al. [15] published results showing that low lighting conditions may reduce the feelings of safety and provide opportunities for vandals. Providing enough light in cities' corridors relaxes the communities and improves surveillance, which in turn reduces vandalism. Herein, the main aim of this study is to explore alternatives of lighting fittings, taking into consideration different wall and flooring materials, in order to achieve acceptable lighting levels for the studied case, which would aid in avoiding any possible vandalism.

1.3. Objectives of the Study

The aim of this study is to discuss how the immature behaviour of a part of the society, resulting in vandalism, affects the building aesthetics and design features in the districts of Abu Shagara, in the Emirate of Sharjah, UAE. The main objectives of this study are:

1. To investigate the main effects of vandalism in the Abu Shagara corridor in relation to lighting.
2. To investigate the effect of lighting and surface materials on the vandalised areas.
3. To investigate the impact of different lightings and different wall and flooring materials on the corridor.

Accomplishing these studies would assist in the development of decisions regarding the methods of improving built environments, especially in the Abu Shagara corridor.

2. Approach and Methodology

The choice of vandalism as a problem to investigate was due to the reason that vandalism is currently common, especially in the areas inhabited by the younger generations. Vandalism plays an important role in the destruction of properties and in street crimes.

2.1. Research Design

Both quantitative and qualitative research approaches were used to evaluate the study outcomes. Various methodology tools were used to collect crucial information for the study. It is important to gain insight into what the public thinks about such an act, and, therefore, surveying people would benefit the study. The methodology tools used were: (1) a questionnaire-based survey, (2) an integrated building performance analysis using the Integrated Environmental Solutions Virtual Environment (IESVE) software package, (3) illuminance measurements. Figures 1 and 2 show the locations of the two sites where the study was conducted.

Figure 1. The Plan of Abu Shagara indicating the two study sites, corridor A and corridor B.

(a)   (b)

Figure 2. (a) Vandalised corridor A showing neighbouring buildings with only windows overlooking the corridor, and (b) non-vandalised corridor B with one neighbouring building (labelled in red) having balconies overlooking the corridor.

The vandalised corridor A (Figure 2a) was compared to the non-vandalised corridor B (Figure 2b); in addition, IESVE simulations were run in order to identify solutions to prevent acts of vandalism in corridor A. In these examples, lighting would prevent vandalism, as suggested by the absence of lighting in corridor A, as can be seen in Figure 2. In this case, a lit-up space was not a comfortable space to perform the act of vandalism.

2.2. Participants in the Questionnaire

In the study, a questionnaire-based survey was conducted aiming to collect crucial information from residents of Abu Shagara, in Sharjah city. The survey included seven questions (Appendix A) that were answered by 44 randomly chosen pedestrians who were roaming around corridor A in Abu Shagara. The objective of the questionnaire was to directly understand the relationship between the respondents' satisfaction with the illuminance levels and their sense of safety. The survey is shown in Appendix A. In November, for a period of two days from Tuesday to Wednesday, two female students collected data around 6:30 p.m., after the lighting was turned on in the corridor. Data collection ended around 8:00 p.m., i.e., when the number of male pedestrians increases for the daily prayer, and before the number of female pedestrians is reduced. The choice of the two corridors A and B in this study was based on their similarity in social aspects and in their layout. The selected corridors A and B are located beside Abu Shagara Park, which is located within a residential area of Sharjah city.

The majority of the respondents were females, with an approximate percentage of 68%. Lorenc et al. [9] stated that increased lighting plays a role in minimizing the fear of criminal actions among pedestrians in residential areas, regardless of their gender and age. The majority of respondents agreed to increase the number of luminaires and the light levels in the corridor to avoid vandalism (score of 4.8) and to facilitate manoeuvres through the corridor (score of 3.43), but reported that they still felt stressed when walking through the corridor (score of 2.89). Also, they showed their dissatisfaction on the present lighting levels, as well as on the visibility in this corridor (scores of 1.57 and 2.25, respectively). Table 1 shows the average response scores for each question of the survey.

Table 1. Average response scores for each question of the survey.

No.	Questions	Av. Score
1	Is the number of Luminaires provided in the corridor sufficient and satisfying?	1.59
2	Do you think the corridor is welcoming in terms of visibility?	2.25
3	Rate the sense of 'safety' when walking through the corridor at night.	3.01
4	Rate the sense of 'stress' when walking through the corridor.	2.89
5	Rate the lighting level in this corridor.	1.57
6	Do you think it is easy to maneuver through the corridor?	3.43
7	Would you agree to increase the number of luminaires and light levels in the corridor to avoid vandalism?	4.80

2.3. Integrated Building Performance Analysis

The IESVE, Integrated Environmental Solutions Virtual Environment, is a program to evaluate and improve buildings in terms of energy, light, ventilation, shade, carbon, lifecycle costs, occupant safety, and economics. We started by drawing the 3D geometry of the existing buildings and corridors and by selecting the city location in the database of IESVE, which takes also climate into consideration when running the software. Second, we selected alternatives for corridors' flooring and wall materials, as well as for electric lamps to achieve better lighting levels in the corridors. The simulation considered the illuminance levels, in lux, as an indicator of the lighting levels. Thirdly, we compared the simulation results for the proposed alternatives among each other and always with the baseline case. Finally, the simulation identified several cases with different inputs that achieved better lighting.

2.4. Illuminance Measurements and Measuring Tape

Illuminance measurements evaluate the illumination of a specific location resulting from the light hitting it, by performing lux measurements [16]. In this study, illuminance measurements were performed by measuring the lux levels in a designated area and by subsequently verifying the results of the IESVE software. The measurements of the lighting differences between the two corridors were our main focus. Further measurements for the vandalised areas in the examined corridor were also performed in this study. At the same time, the illuminance measurements were used to calculate the lux levels in the corridors in order to compare the differences in lux levels. Corridor A was vandalised in two wall areas (Figure 3). The points labelled in the corridor represent the maximum distance for a visitor to vandalise a wall and be seen by a passing visitor (Figure 4). The shaded area is the area vandalised in the corridor. Corridor A consists of two vandalised areas, one area of 2.85 m^2 and the other of 3.3 m^2. Therefore, the total area vandalised in the corridor is of 6.15 m^2. The lux level of each vandalised area was measured in the middle using the illuminance measurements. Four reading trials were conducted for each area, and the average lux level of points A and B were approximated to 2.95 and 3.18 lux, respectively.

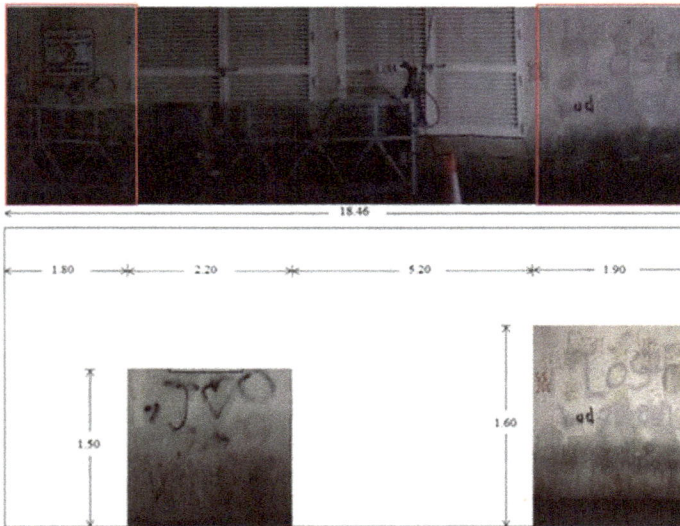

Figure 3. Image illustrating the width and height in meters of the vandalised wall in corridor A.

Corridor B is not affected by vandalism; points for measurements were established in the middle of the corridor based on the lamp positions. The lux level was calculated at each point in four readings to calculate the average lux values (Figure 5). Readings of A, B, C, and D were 33.28, 44.83, 31.55, and 28.9, respectively. Because this corridor receives light from the surrounding shops, no vandalised areas are seen in the corridor, differently from corridor A.

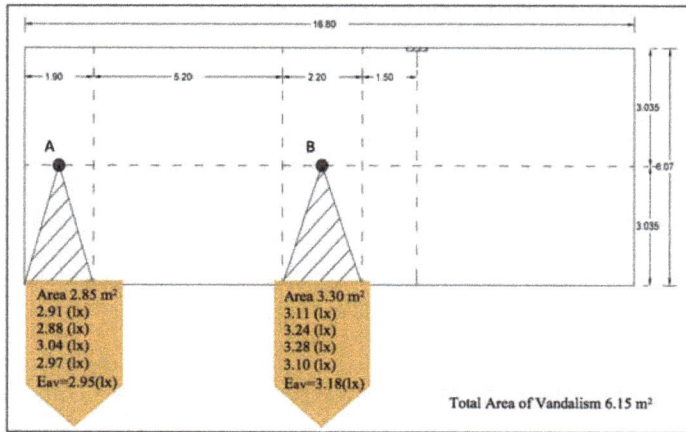

Figure 4. Corridor A plan illustrating the vandalised area and the amount of lux (lx) available at each vandalised point, with the illuminance measured at a mid-distance in the corridor.

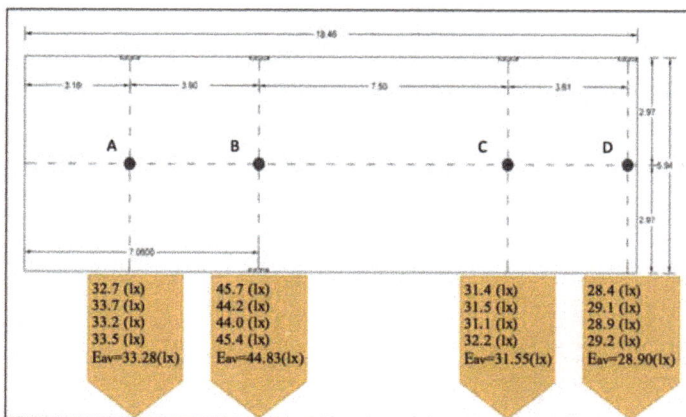

Figure 5. Corridor B plan illustrating the amount of lux (lx) at each point where the lamps are located, with the illuminance measured at a mid-distance in the corridor.

3. Investigated Area and Analysis

3D modelling of the investigated area of Abu Shagara, which considered the current conditions of the two corridors A and B, was created through the IESVE software (Figure 6). After running the simulation, the results showed the same lux levels measured on site using illuminance measurements (Figure 4). Moreover, the simulation showed lux levels lower than 3 lux (Figure 4), i.e., lower than the local standards explained in Section 1.1. This led to vandalism in corridor A, encouraged by the insufficient illuminance, as opposed to corridor B that is exposed to sufficient illuminance. The horizontal illuminance is the criterion determining the safety of movement for pedestrians on the road, and the vertical illuminance and semi-cylindrical illuminance are the criteria governing the ability of facial recognition [11]. Therefore, vandals can pass through a space lacking sufficient illuminance and vandalize community assets, for example with graffiti.

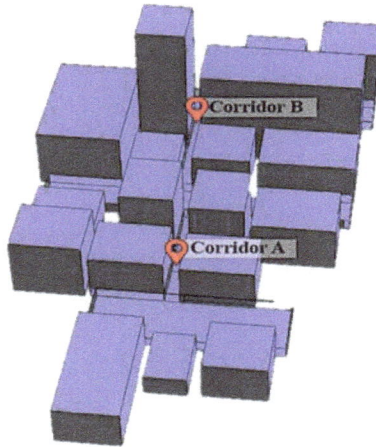

Figure 6. 3D view of the Abu Shagara site obtained with the IESVE software package, illustrating the locations of corridors A and B.

Corridor A—Vandalised

The current conditions of corridor A were recreated on the IESVE software. A simulation was conducted using the same lux levels measured on the site using illuminance measurements, as illustrated in Figure 4. The simulation showed lux levels lower than 3 lux (Figure 7). Likewise, the current conditions of corridor B were recreated on the IESVE software package following the same lamp design distribution as that on the site, as shown in Figure 5. The lux levels of corridor B, determined by the simulation, were similar those measured on the site, i.e., ranging between 25 to 45 lux (Figure 8).

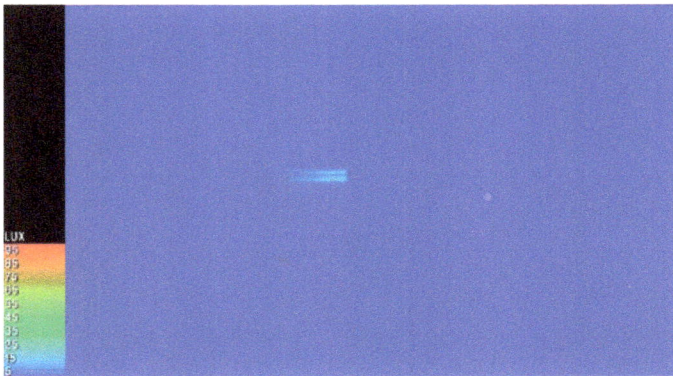

Figure 7. IESVE illuminance simulation for corridor A in the current situation, illustrating lux levels below 3 lx.

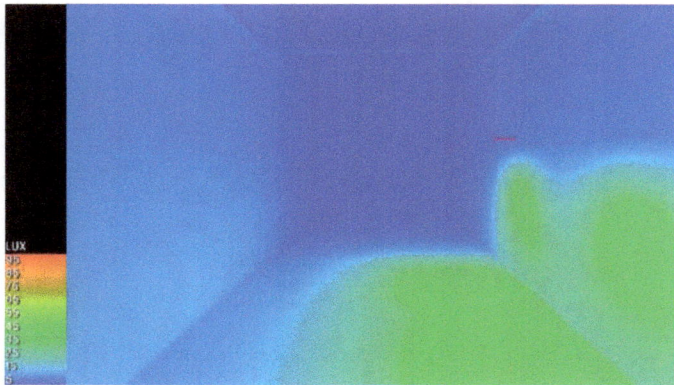

Figure 8. IESVEVE illuminance simulation for corridor B in the current situation, illustrating lux levels ranging between 25 and 45 lx.

4. Simulation Cases and Results

Ten cases were analysed based on the comparison between the two corridors and the IESVE simulation of their current situation in terms of luminance. Table 2 illustrates the first four cases and the parameters covered in the simulation, as well as the materials used for parameters 2 and 3.

Table 2. Parameters (P) and types of lighting in the different simulation cases.

Cases		P1: Corridor B—Lighting	P2: Flooring Material		P3: Wall Material	
Case 1		√				
Case 2		√	√			
Case 3		√			√	
Case 4		√	√		√	
Types of lighting						
	Light Type	Manufacturer	Flux (lm)	Width (m)		Height (m)
Case 5	L1: Linear	ABS Lighting	5600	0.19		0.01
Case 6	L2: Point	Fluora	1200	0.19		0.01
Case 7	L3: Linear	Spectral	4650	0.13		0.03
Case 8	L4: Point	Hess	2400	0.26		0.5
Case 9	L5: Point	iGuzzini	950	0.18		0.08
Case 10	L6: Point	Targetti-exterieur vert	950	0.17		0.10
Material selection for walls and floorings						
Type		Material				
Walls		Shining painting wall				
Floors		Polished concrete floor				

4.1. Case 1—Lighting Installation

This simulation consisted in installing lightings in corridor A based on the lightings used in corridor B and in analysing their impact on the corridor, the illuminance distribution, and the lux levels. The illuminance simulation showed that the resulting lux levels ranged from 25 to 45, differing from one wall of the corridor to the other, as illustrated in Figures 9 and 10. This huge increase in corridor A lighting showed how corridor A is uninviting in its current illumination conditions. The exposure of corridor A to the street increased when the lux levels in it increased, which would aid in avoiding vandalism.

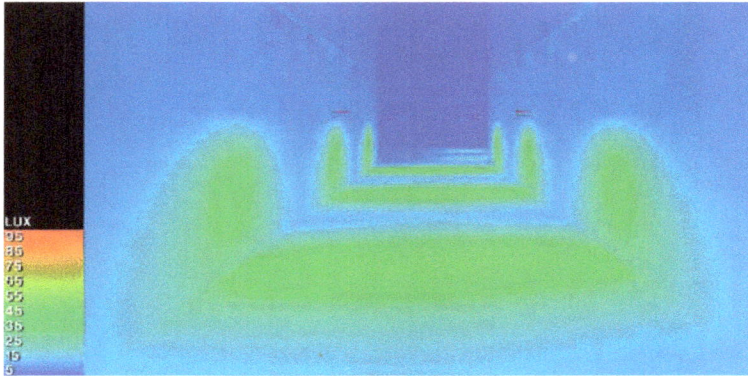

Figure 9. Case 1: illuminance simulation and lux distribution.

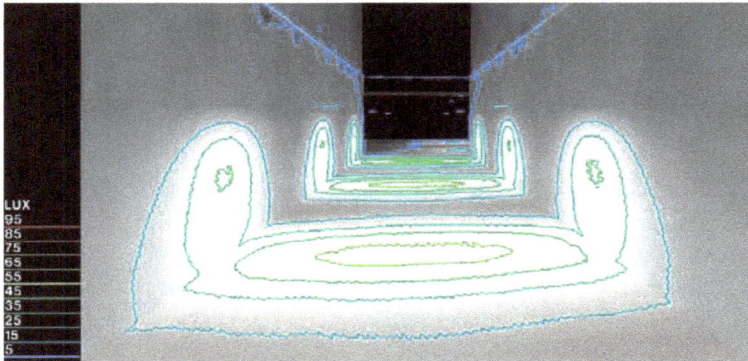

Figure 10. Case 1: illuminance simulation and lux contour distribution.

4.2. Case 2—Flooring Material and Lighting Installation

This simulation consisted in installing in corridor A flooring material and lightings similar to those used in corridor B and in testing the effects of the flooring material and the lighting type on illuminance. The illuminance simulation showed that the resulting lux level ranged from 35 to 55, varying from one wall to the other, as illustrated in Figures 11 and 12.

These results showed that higher lux levels were achieved compared to case 1, which, therefore, indicate that changing the flooring material along with adding lighting fixtures enhance the corridor's exposure to the street.

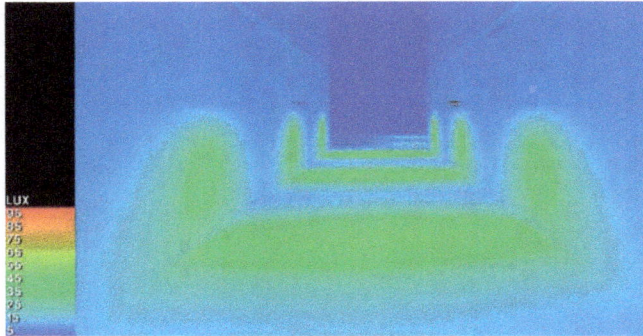

Figure 11. Case 2: illuminance simulation and lux distribution.

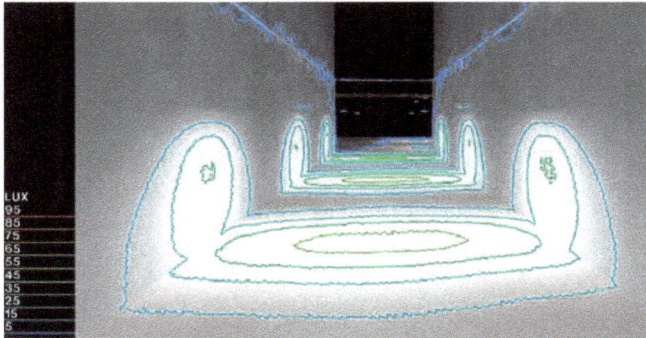

Figure 12. Case 2: illuminance simulation and lux contour distribution.

4.3. Case 3—Wall Material and Lighting Installation

This simulation consisted in installing in corridor A wall materials and fluorescent lightings similar to those present in corridor B. The analysis of the impact of the wall materials and the lighting on corridor A illuminance showed that the resulting lux level ranged from 35 to 55, varying from one wall to the other, as illustrated in Figures 13 and 14.

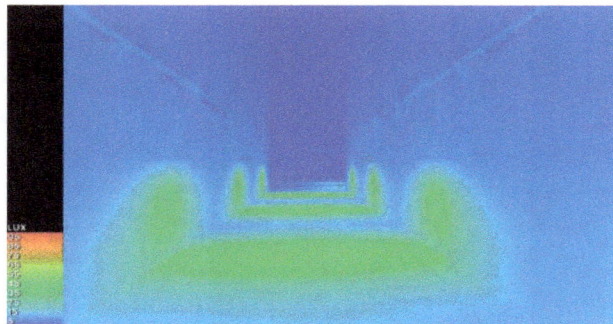

Figure 13. Case 3: illuminance simulation and lux distribution.

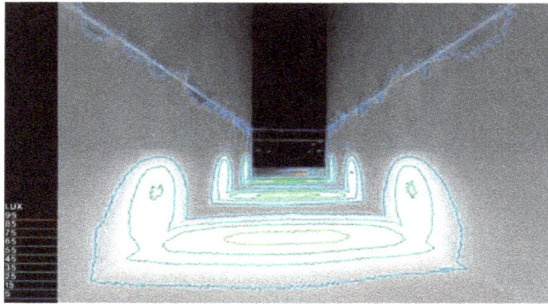

Figure 14. Case 3: illuminance simulation and lux contour distribution.

4.4. Case 4—Wall Material, Flooring Material and Lighting Installation

This simulation consisted in installing in corridor A wall and flooring materials, as well as fluorescent lightings similar to those present in corridor B and in analysing their impact on corridor A illuminance. The illuminance simulation showed that the resulting lux levels ranged from 35 to 55, varying from one wall to the other, as illustrated in Figures 15 and 16.

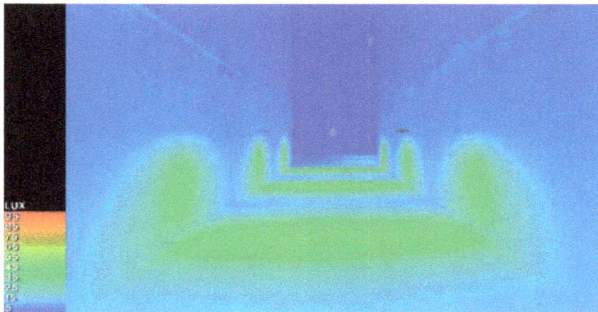

Figure 15. Case 4: illuminance simulation and lux distribution.

Figure 16. Case 4: illuminance simulation and lux contour distribution.

By comparing the results of the four simulations, it was found that the impact of the different parameters on the corridor illuminance differed slightly in lux levels. Therefore, when combining cases 1, 2, and 3, as it was done in case 4, the resulting impact on illuminance was slightly different compared to that measured in each of the single cases, as shown in Figures 15 and 16 for case 4. It was shown that the lighting had a greater impact on the lux levels and lux distribution than the surface materials. For this reason, different IESVE illuminance simulation analyses on different lightings were conducted in order to determine the most convenient lighting type in terms of lux levels and lux distribution (Table 2). Figures 17–22 illustrate the photometric polar diagrams of the lighting types L1, L2, L3, L4, L5, and L6, for which IESVE illuminance simulations were performed.

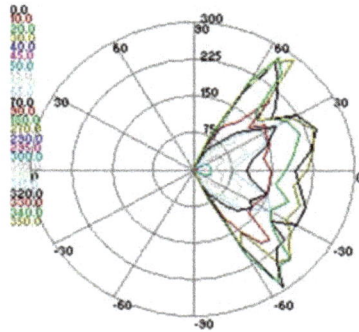

Figure 17. L1: Linear ABS lighting.

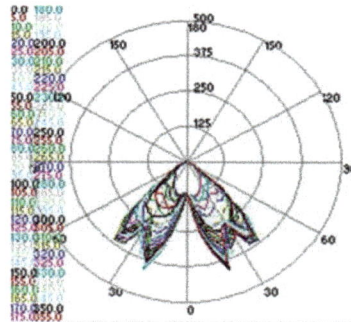

Figure 18. L2: Point Fluora.

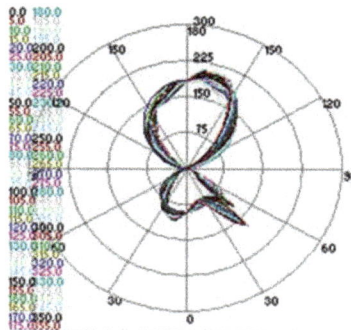

Figure 19. L3: Linear Spectral.

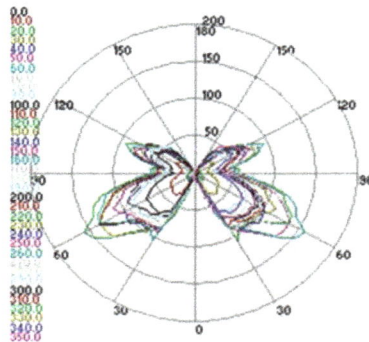

Figure 20. L4: Point Hess.

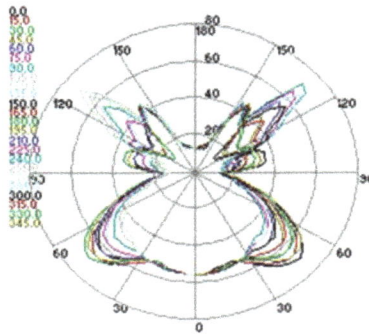

Figure 21. L5: Point iGuzzini.

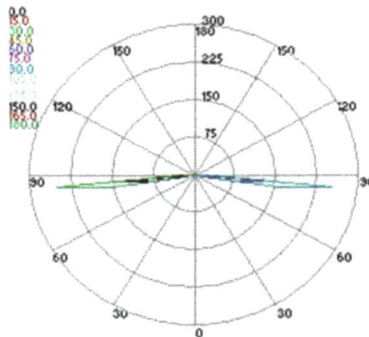

Figure 22. L6: Point Targetti-exterieur vert.

4.5. Case 5—Lighting Type L1: Linear ABS lighting

The illuminance simulation in the presence of the lighting type L1 showed that the lux levels reached values in the range from 75 to 90 lux. The light was mostly distributed on the ground and at the lower extremities of the walls, as illustrated in Figures 23 and 24. Case 5 showed higher lux levels than those reached in cases 1 to 4, which, in addition, increased the street exposure of corridor A.

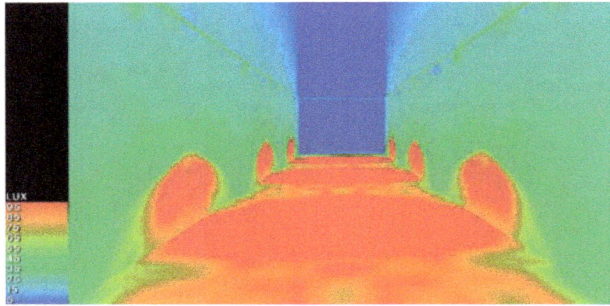

Figure 23. Case 5: illuminance simulation and lux distribution.

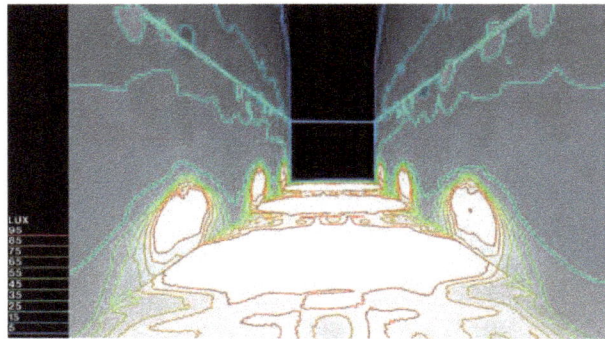

Figure 24. Case 5: illuminance simulation and lux contour distribution.

4.6. Case 6—Lighting Type L2: Point Fluora

The illuminance simulation in the presence of the lighting type L2 produced lux levels in the range from 45 to 55 lux, with levels ranging from 65 to 75 lux in the middle of the corridor. The light was distributed in a narrow area on the floor and along the lower extremities of the wall, as illustrated in Figures 25 and 26. Case 5 showed higher lux levels than cases 1–4, whereas, when installing lighting type L2, as in case 6, the lux level were not as high as in case 5.

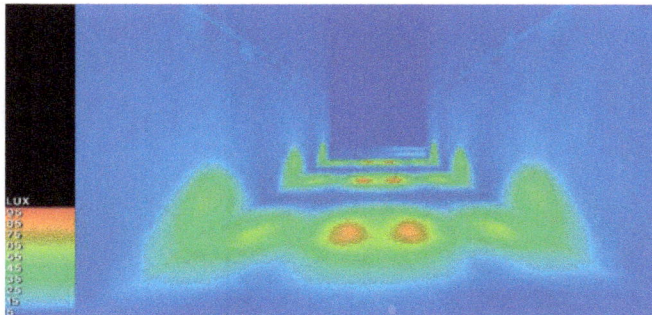

Figure 25. Case 6: illuminance simulation and lux distribution.

Figure 26. Case 6: illuminance simulation and lux contour distribution.

4.7. Case 7—Lighting Type L3: Linear Spectral

The illuminance simulation with the lighting type L3 showed that the lux level reached values ranging from 35 to 55 lux, distributing on the top and bottom of the wall surface, and values ranging from 85 to 90 lux in the middle area, as illustrated in Figures 27 and 28. Case 7 showed a higher lux level than any of the previous six cases; therefore, the lighting type L3 resulted particularly effective in keeping the corridor A highly exposed to the street.

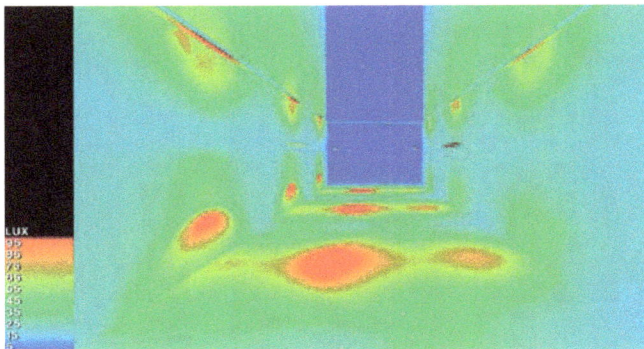

Figure 27. Case 7: illuminance simulation and lux distribution.

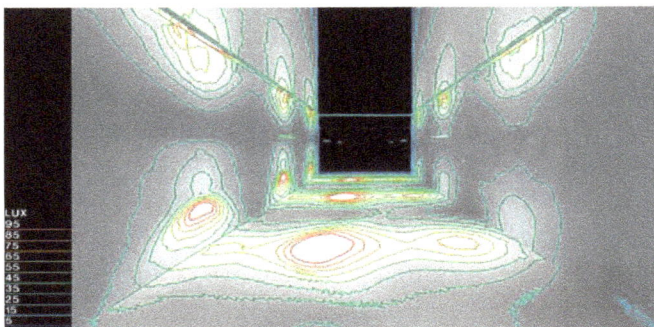

Figure 28. Case 7: illuminance simulation and lux contour distribution.

4.8. Case 8—Lighting Type L4: Point Hess

The illuminance simulation with the lighting type L4 showed that the lux levels reached values in the range from 35 to 55 lux, with the light distributed vertically over a large area of the wall, which plays well against vandalism because walls are usually vandalised. The lux levels were also within the same range observed for corridor B. In addition, the lux levels reached values ranging from 75 to 85 lux in a small area on the wall, as illustrated in Figures 29 and 30. Case 8 showed similar lux levels as case 5, and the walls were lighted up to a certain height with high lux levels that were better distributed than case 7, which had higher lux levels than case 8, but the light was fragmented on the walls that are subject to vandalism.

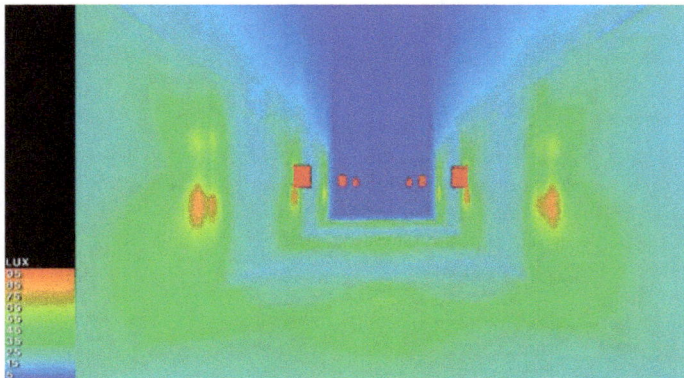

Figure 29. Case 8: illuminance simulation and false colours in lux.

Figure 30. Case 8: illuminance simulation and false lines in lux.

4.9. Case 9—Lighting Type L5: Point iGuzzini

The illuminance simulation with the lighting type L5 showed that the lux levels reached values ranging from 5 to 25 lux, showing a weak distribution of light over the walls, as illustrated in Figures 31 and 32. Case 9 showed the lowest lux levels among all cases; therefore, case 8 was the best choice.

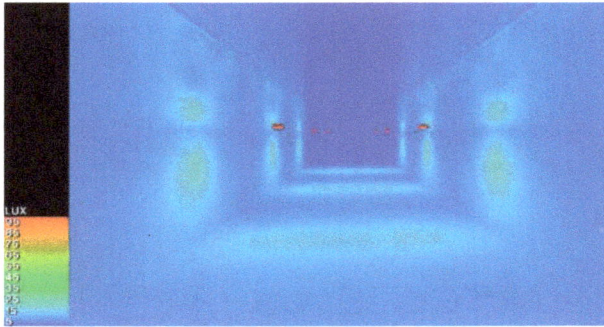

Figure 31. Case 9: illuminance simulation and false colours in lux.

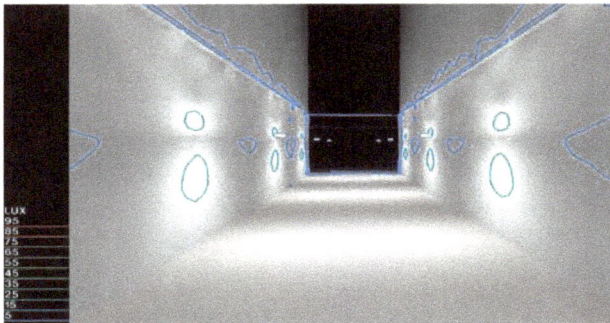

Figure 32. Case 9: illuminance simulation and false lines in lux.

4.10. Case 10—Lighting Type L6: Point Targetti-Exterieur Vert

The illuminance simulation with the lighting type L6 showed that the lux levels reached values ranging from 5 to 15 lux, with a small vertical area showing values ranging from 45 to 55 lux. As shown in Figures 33 and 34, there was a weak distribution of light over a small area on the walls, and the lux levels were low compared to corridor B. Case 10 showed similar lux levels as case 9, thus both cases showed poor lux levels, whereas case 8 showed the highest lux levels.

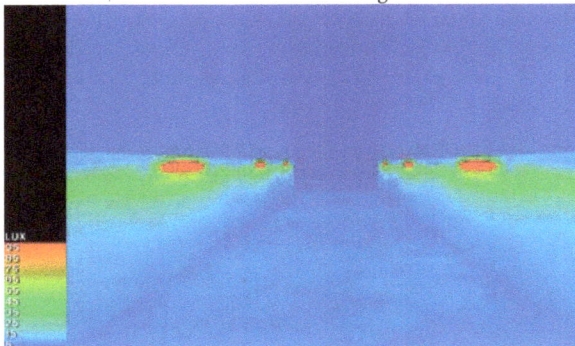

Figure 33. Case 10: illuminance simulation and false colours.

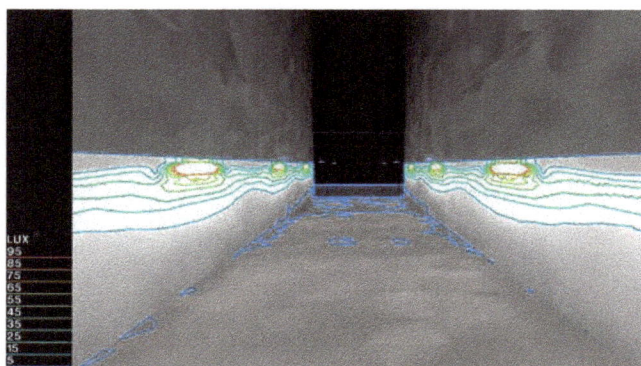

Figure 34. Case 10: illuminance simulation and false lines.

5. Conclusions

In this study, the impact of lighting on vandalism in hot climates was examined by considering the case of the Abu Shagara vandalised corridor in Sharjah, UAE. After studying the cases 1 to 4, the simulation results led to the conclusions that the type of wall and flooring materials had a slight impact on the illuminance levels and distribution compared to the lighting fixtures. In all four cases, the lighting fixtures had a greater impact on the lux levels of corridor A, increasing the light intensity from, approximately, 2 to 3 lux to 25 to 45 lux. There is a correlation between the type of light fixture and vandalism activity, as observed through different comparisons. These results led to the study of the impact of six types of light lamps on corridor A. A study of the photometric polar diagrams of the lighting types L1, L2, L3, L4, L5, and L6, as well as of their light distribution was conducted. Based on the simulation analysis, case 8, with the lighting type L4, was found to provide the most suitable lighting type for corridor A, among all cases examined (case 5 to case 10). The illuminance simulation in case 8 showed that the lux levels reached values ranging from 35 to 85 lux, and the illuminance was distributed over a large vertical area on the wall around the light fixtures; in addition, lux levels ranging from 25 to 35 lux were observed covering vertical areas on the wall between the light fixtures. Thus, the lighting type L4 can illuminate the walls up to the point reached by vandalism and even beyond, which reduces the possibilities of vandalism. On the other hand, the lux levels reached values ranging from 25 to 55 lux in horizontal areas in the corridor. Herein, the national and international standards discussed earlier were successfully achieved. This study could assist architects and property owners in the prevention of vandalism.

Moreover, since this study was conducted in the hot climatic conditions of the UAE, the impact of lighting on vandalism could be also examined in other climatic conditions in different areas, to effectively evaluate the results in different situations. In general, this approach is distinctive and applicable to other countries and areas around the world.

Author Contributions: All authors contributed to this work. Emad Mushtaha generated the idea and designed the experiment as well as the simulation scenarios. Ranime Ayssar Nahlé and Maitha Bin Saifan prepared and analysed the questionnaire as well as performed the experiments and simulations. All authors, including Hasim Altan prepared, revised, and approved the final manuscript.

Conflicts of Interest: the authors declare no conflict of interest.

Appendix A

PERSONAL DATA

AGE:

GENDER:　☐ Male　　☐ Female

We would like to receive your input regarding the lighting in this vandalized corridor. Please fill in this anonymous survey where 1 is the minimum and 5 is the maximum.

	1	2	3	4	5
1. Is the number of Luminaires provided in the corridor sufficient and satisfying?					
2. Do you think the corridor is welcoming in terms of visibility?					
3. Rate the sense of 'Safety' when walking through the corridor at night.					
4. Rate the sense of 'Stress' when walking through the corridor.					
5. Rate the lighting level in this corridor					
6. Do you think it is easy to maneuver through the corridor?					
7. Would you agree to increase the number of luminaires and light levels in the corridor to avoid vandalism?					

THANK YOU FOR YOUR COOPERATION

References

1. Miller, A. *Vandalism and the Architect*; Ward, C., Ed.; Architectural Press: London, UK, 1973.
2. Teng, H.; Puli, A.; Karakouzian, M.; Xu, X. Identification of Graffiti Countermeasures for Highway Facilities. *Procedia Soc. Behav. Sci.* **2012**, *43*, 681–691. [CrossRef]
3. Mushtaha, E.; Hamid, F. The Effect on Vandalism of Perception Factors Related to Housing Design, Case of UAE Cities. *J. Asian Archit. Build. Eng.* **2016**, *15*, 247–254. [CrossRef]
4. Ley, D.; Cybriwsky, R. Urban Graffiti as Territorial Markers. *Ann. Assoc. Am. Geogr.* **1974**, *64*, 491–505. [CrossRef]
5. Christensen, H.H.; Johnson, D.R.; Brookes, M.H. *Vandalism: Research, Prevention, and Social Policy*; Gen. Tech. Rep. PNW-GTR-293; U.S. Department of Agriculture, Forest Service, Pacific Northwest Research Station: Portland, OR, USA, 1992.
6. Cohen, S. Hooligans, Vandals and the Community: A Study of Social Reaction to Juvenile Delinquency. Ph.D. Dissertation, London School of Economics and Political Science (LSE), London, UK, 1969.
7. Peña-García, A.; Hurtado, A.; Aguilar-Luzón, M. Impact of public lighting on pedestrians' perception of safety and well-being. *Saf. Sci.* **2015**, *78*, 142–148. [CrossRef]
8. Abd-Razak, M.; Mustafa, N.; Che-Ani, A.; Abdullah, N.; Mohd-Nor, M. Campus sustainability: Student's Perception on Campus Physical Development Planning in Malaysia. *Procedia Eng.* **2011**, *20*, 230–237. [CrossRef]
9. Lorenc, T.; Petticrew, M.; Whitehead, M.; Neary, D.; Clayton, S.; Wright, K.; Renton, A. Fear of crime and the environment: Systematic review of UK qualitative evidence. *BMC Public Health* **2013**, *13*, 496. [CrossRef] [PubMed]
10. ARUP. The Value of Public Spaces: Economic or Social? ARUP, p. 1. Retrieved from Frederikssund Kommune. Available online: www.frederikssund.dk (accessed on 9 March 2015).
11. Kim, D.; Park, S. Improving community street lighting using CPTED: A case study of three communities in Korea. *Sustain. Cities Soc.* **2017**, *28*, 233–241. [CrossRef]
12. Department of Municipal Affairs Abu Dhabi. *Abu Dhabi Public Realm & Street Lighting Handbook*, 1st ed.; Department of Municipal Affairs Abu Dhabi: Abu Dhabi, UAE, 2014.
13. CIE. *Lighting of Roads for Motor and Pedestrian Traffic*; International Commission of Illumination, CIE Public: Vienna, Austria, 2010; Volume 115.
14. Farrington, D.; Welsh, B. Measuring the Effects of Improved Street Lighting on Crime: A reply to Dr. Marchant 1. *Br. J. Criminol.* **2004**, *44*, 448–467. [CrossRef]

15. Blöbaum, A.; Hunecke, M. Perceived danger in urban public space: The impacts of physical features and personal factors. *Environ. Behav.* **2005**, *37*, 465–486. [CrossRef]
16. Cheng, W.; Hsu, C.; Chao, C. Temporal vision-guided energy minimization for portable displays. In Proceedings of the 2006 International Symposium on Low Power Electronics and Design, Tegernsee, Germany, 4–6 October 2006; pp. 89–94.

Article

Structuring the Environmental Experience Design Research Framework through Selected Aged Care Facility Data Analyses in Victoria

Nan Ma [1], Hing-wah Chau [1], Jin Zhou [2] and Masa Noguchi [1,*]

[1] ZEMCH Lab, Faculty of Architecture, Building and Planning, The University of Melbourne,
     Parkville, VIC 3010, Australia; nanm@student.unimelb.edu.au (N.M.); chauh@unimelb.edu.au (H.-w.C.)
[2] Thrive Research Hub, Faculty of Architecture, Building and Planning,
     The University of Melbourne Parkville, VIC 3010, Australia; jin.zhou@unimelb.edu.au
*   Correspondence: masa.noguchi@unimelb.edu.au; Tel.: +61-3-9035-8193

Received: 4 September 2017; Accepted: 23 November 2017; Published: 25 November 2017

Abstract: Humans relate to the living environment physically and psychologically. Environmental psychology has a rich developed history while experience design emerged recently in the industrial design domain. Nonetheless, these approaches have barely been merged, understood or implemented in architectural design practices. This study explored the correlation between experience design and environmental psychology. Moreover, it conducted literature reviews on theories about emotion, user experience design, experience design and environmental psychology, followed by the analyses of spatial settings and environmental quality data of a selected aged care facility in Victoria, Australia, as a case study. Accordingly, this study led to proposing a research framework on environmental experience design (EXD). It can be defined as a deliberate attempt that affiliates experience design and environmental psychology with creation of the built environment that should accommodate user needs and demands. The EXD research framework proposed in this study was tailored for transforming related design functions into the solutions that contribute to improving the built environment for user health and wellbeing.

Keywords: environmental experience design; user experience design; function analysis; environmental psychology; indoor environmental quality; aged care facilities; architectural design; EXD research framework matrix

1. Introduction

Architectural design encompasses a complex thought process that leads to realising the configurations and spaces with due consideration of market needs and demands. The growth of ageing population is on the hike. However, many existing aged care facilities today barely accommodate the concept of age-friendliness. The personalisation in the design development may need to be taken into account. Design decision makers, such as architects, are responsible for creating the physical and psychological comfortable built environment. Psychological comfort is subject to human emotion. Crozier defines emotion as "conceptions of meaningful responses to life experience" [1]. He also mentions that "it is a person's experience of the world rather than the world's objective properties that counts" [1]. Furthermore, McCarthy and Wright annotate that "emotions are qualities of particular experiences" [2,3]. In some manners, these statements support the importance of design for built environment that needs to accommodate proper user experiences.

It is worth noting that the "experience design" has been devised in the industrial design domain, originated from "user experience design" which focuses on the design of end-user product interfaces [4]. Experience design aims to achieve physical and psychological spatial needs and demands from user perspectives [4]. The term of user experience design was coined by Norman attempting to refine the interface between humans and facilities in order to emphasise the essentials of human interaction design [5]. A synergic theory expounded by O'Sullivan and Spangler is concerned with involvement and creation of an environment establishing connections on the level of emotion or value to users [6]. "It is now about creating experiences beyond just products and services, about creating relationships with individuals, creating an environment that connects on an emotional or value level to the customer" [4]. Pine and Gilmore's statement clarifies that the most successful experience design may have "sweet spots" regarding both passive and active participation of end-users and absorptive and immersive connection in its consistent context [7]. Hassenzah describes that "experience design ... requires a broadened perspective, with the fulfillment of psychological needs (values), which in turn creates meaning and emotion, as the prime design objective" [4]. Berridge defines experience design as "a new emerging paradigm, a call for inclusion: it calls for an integrative practice of design" [8]. It is a new term, making an effort on exploring the connection between the concept of experience and the practice of design. However, "the design of experience is not any newer than the recognition of experience ... it is really combination of many previous disciplines" [8].

Architectural design involves interdisciplinary and human-centric decision-making processes; however, experience design has barely been applied to architectural practices. Kellert et al. indicate that the sense of gratification within a building is "an experience of architectural pleasure that resonated as both new and unfathomably familiar" [9]. The ambition of experience design is to orchestrate human experiences with functionality, purposefulness, engagement, stimulation and memory [10]. Experience design applied to the built environment may bring up potentials to improve the understanding of association between residents and their physical settings. Exploring experience design, when implemented in the built environment, articulates a paradigm which emphasises the necessity of enhancing and restoring human experiences [11]. In this respect, this application may help replace the technology-driven decision-making approach by the human-centric one. Frumkin provides empirical evidences on how humans respond psychologically to the built environment [12]. It may acknowledge that the research on environmental psychology provides insights on how human-centric experience design engages with architectural practices. Gifford defines environmental psychology as "a study of transactions between individuals and their physical settings" [13]. Furthermore, Nagar also proposes a similar definition, arguing environmental psychology is "the study of the interrelationships between the physical environment and human behaviour" [14].

Human emotion to some extent correlates to user experience; architecturally, it may contribute to generating the notion of experience design linked to environmental psychology (Figure 1). This paper aims to develop knowledge of the experience design and to propose a research framework that enhances the interrelationship between residents and the built environment. It uses a selected aged care facility in Victoria, Australia, as a case study to document and analyse the spatial settings and physical environmental conditions. The following sections identify the present circumstances for further analyses.

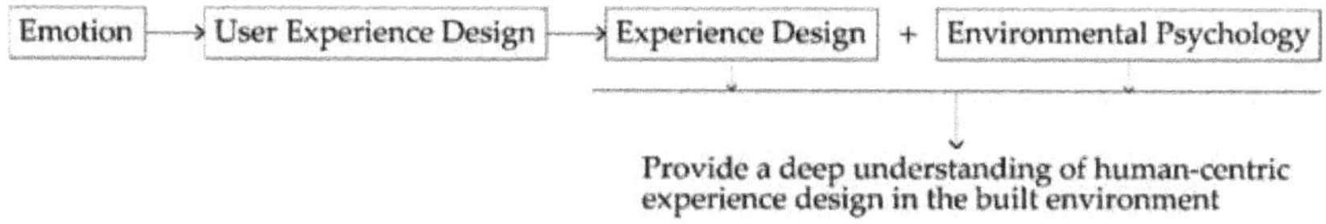

Figure 1. Structure of the literature review.

2. Review on Residential Aged Care Facilities in Australia

Residential aged care facilities cater for the elderly who cannot live independently in their own homes. These facilities were designed as a "miniaturised acute-care hospital" in the post-war era [15]. In 1980s, as reflected in the *Guidelines for the Provision of Nursing Home Facilities* published by the Australian government in 1983, residents were named as "patients" and their rooms were defined as "wards". Multi-bedded rooms were recommended with shared toilets and bathrooms. Operational efficiency and effective surveillance were promoted instead of personal privacy and dignity [16]. Collective living and rigid routines affected the individuality and personal identity of residents [17]. The institutionality of residential aged care facilities was criticised by Carboni as the "negation of home" with the "sense of placelessness" due to the loss of intimacy between the individual and the environment [18]. According to Maslow, there are at least five sets of human needs, ranging physiological, safety, love/belonging, esteem and self-actualisation [19]. Apart from the fundamental physical needs, personal safety and security, it is crucial for people to have the sense of belonging and to be accepted and respected by others to build up their self-esteem and confidence against loneliness, isolation and depression. Due to various impairments, it is common for the elderly to experience stress and frustration, so they are vulnerable to environmental impacts [20]. The design of aged care facilities should not merely provide the residents a comfortable and safe place to live, but also facilitate social interaction among them and enable them to exercise their residual abilities towards self-actualisation without much difficulty.

Individual needs of the elderly were emphasised by Kidd in *The Image of Home: Alternative Design for Nursing Homes*, published by the Centre for Applied Research on the Future at the University of Melbourne in 1987. According to Kidd, "every effort should be made to restore, retain, regain or develop independence, choice and decision making" amongst residents [21]. The physical setting may need to offer a variety of communal areas for interacting with others and personalisation of private spaces according to individual preferences. Similar ideas were incorporated by Kidd in the subsequent *Hostel Design Guidelines* published by the Commonwealth Department of Community Services and Health in 1988. Based on this guideline, residential aged care facilities should reflect a truly domestic character to avoid the risk of disorientation within the buildings and to provide a stimulating environment for both social encounters and personal retreat. The importance of single bedroom provision was also emphasised for enabling residents to personalise their individual accommodation [22]. The proportion of single bedroom in a residential aged care facility has been significantly increased from one per ten residents according to the guidelines in 1983 to at least 75% according to the *Aged Care Residential Services Generic Brief* published by the Victorian Government in 2000 [23].

3. The Case Study

In Victoria, there are 750 government-funded nursing homes and 131 privately operated supported residential services (SRS) [24,25]. Most of the facilities cater for English-speaking residents or a specific cultural group. Adare SRS is unique in terms of having residents with multi-cultural backgrounds, so it was selected as the site for investigation.

3.1. Spatial Design Setting

Adare SRS is located in Wantirna South, approximately 35 km to the south-east of Melbourne's Central Business District. It was built in the year of 2000 and can accommodate 45 residents. As of August 2017, about one-third of residents are English-speaking, while the remaining speak Cantonese, Mandarin, Shanghainese and other Chinese dialects. Room numbers are labelled from 1 to 39, 60 to 63, 65 and 66 to avoid unlucky numbers according to Chinese traditional belief. Apart from Rooms 16 and 17 which are connected together to accommodate an aged couple, the remainders are single bedrooms with ensuite (Figure 2). Bedrooms are located on both sides of three major corridors. The longer corridor (Corridor 1: Rooms 1 to 21) and the shorter corridor (Corridor 2: Rooms 22 to 31) run north–south, whereas the third corridor (Corridor 3: Rooms 32 to 66, together with kitchen and laundry) run east–west. The communal spaces (living lounges and dining areas) link these three corridors together with gardens on both sides of the living lounges.

Compared with the *Aged Care Residential Services Generic Brief* published in 2000, Adare SRS has a very high percentage of singe bedrooms, which can provide privacy for each resident. Residents are allowed and even encouraged to bring their own furniture and decorative fixtures to their individual rooms which can foster their sense of identity (Figure 3). The inclusion of personal fixtures and items in the rooms, including chairs, photos, paintings and other artefacts, can reinforce residents' residual memory capacities and arouse their reflection upon their past experiences. This is helpful to personalise the institutional setting and create a familiar environment for them to live. The display of personal objects may stimulate social interaction and conversation among residents and may enable the staff to improve the understanding of the residents about their stories and preferences.

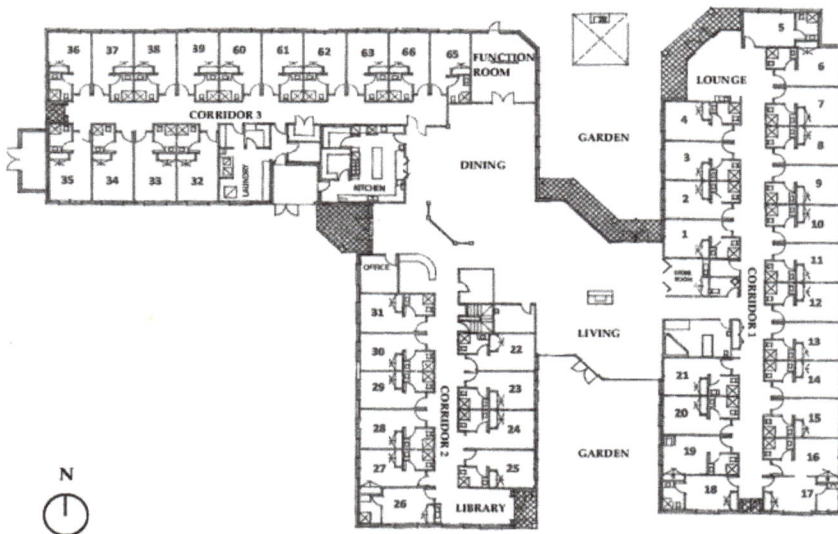

Figure 2. Floor plan of Adare supported residential services (SRS).

Figure 3. Residents' rooms with personal furniture and decorative fixtures.

Various communal spaces are provided at Adare SRS to cater for different types of social encounters among residents depending on their choices. Couches in the living area are arranged in clusters and small family-sized tables for four to six residents are provided in dining area (Figure 4). Besides living and dining area, there are a small library and a small lounge at the end of Corridors 2 and 3, respectively (Figure 5, Left Panel). A private function room is provided next to the dining area for family gathering or other activities which can be booked in advance by residents and their family members to suit their needs (Figure 5, Right Panel).

Figure 4. Living area (**left**); and dining area (**right**).

Figure 5. Lounge (**left**); and private function room (**right**).

Outdoor gardens are located on both sides of the communal living area (Figure 6). Doors opening to the gardens are unlocked during the daytime which can facilitate residents to go outside as one of their choices. In fact, a well-designed garden is a therapeutic environment for the elderly as it can provide visual, tactile, olfactory and auditory stimulation through the combination of natural landscape, fragrance, sunlight, wind and birds.

Figure 6. Outdoor gardens.

There is no air-conditioner at Adare SRS. During summer time, evaporative coolers at communal areas can be switched on and there is a ceiling fan in each resident's room for cooling purpose. During winter time, wall-mounted hydraulic heaters at both communal areas and residents' rooms can be turned on to keep the interior space warm. Residents can open windows in their rooms for natural ventilation and adjust window curtains to control the amount of sunlight into their rooms.

3.2. Indoor Environmental Quality Data Collection and Analysis

Indoor environmental quality measurements were conducted over a one-week period from 29 May to 4 June 2017. For the analysis in this paper, we consider the data collected on five weekdays. The collected data can represent conditions in winter season because the three most important factors that influencing indoor environmental quality—climate condition, elderly activity pattern, and building operation, differs little across winter days. The investigation focusing on summer season is warranted in future studies. Sampling was performed with 2-min time resolution. Two sets of instruments were concurrently deployed at Adare SRS. One set of instruments was located in Bedroom 21 and the second monitoring station was operated in the dining area. The dining hall is selected as the sampling location for public spaces because it is the most used communal area and it share space with living area.

Real-time, size-resolved optical particle counters (OPC, model 9306, TSI Inc., Shoreview, MN, USA) were employed to measure concentrations of airborne particles at Adare SRS. These particle counters have a detectable minimum particle diameter of 0.3 μm. Number concentrations were reported in six size bins based on optical diameter (0.3–0.5, 0.5–1.0, 1.0–2.5, 2.5–5.0, 5.0–10.0, and >10 μm). Particle sampling was performed with 2-min time resolution. This study focuses on the data covering the diameter range 0.3–2.5 μm ($PM_{0.3-2.5}$) and 0.3–10 μm ($PM_{0.3-10.0}$).

Real-time measurements of dry bulb temperature, radiant temperature, relative humidity, and carbon dioxide (CO_2) concentration were made by means of indoor air quality monitor (HuxConnect, Hux Pty. Ltd., Melbourne, Australia). Measurements were recorded with 5-min time resolution.

In addition to the indoor measurements, one additional set of instruments was operated at a local outdoor monitoring site to record the outdoor particle levels, dry bulb temperature, and relative humidity on a continuous basis. Over 30,000 data points were generated from the one-week monitoring campaign. For each indoor air quality index of interested, the distribution patterns were comparable across monitoring days.

(1) CO_2 concentrations

The elevated concentration of CO_2 has been associated with negative impacts on human cognition and decision-making [26,27]. ANSI/ASHRAE Standard 62.1-2016 recommends that indoor CO_2 concentration should be below 1000 ppm [28]. As illustrated in Figure 7, The CO_2 concentrations measured at both bedroom and communal space were kept below 1000 ppm throughout the whole monitoring period. It is worth noting that the 1000 ppm is the concentration threshold preferred by healthy adults. The guideline on CO_2 exposure limit for elderly group has not been established.

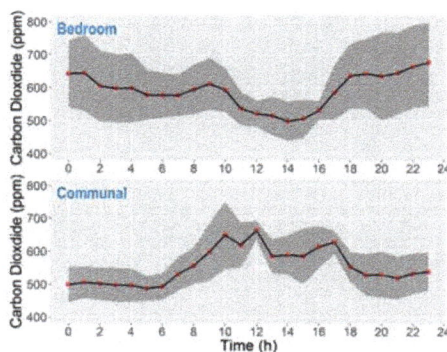

Figure 7. Daily 1-h mean (±standard error) concentrations of carbon dioxide measured at bedroom and communal space over the full monitoring period. The shaded areas represent variability of results as ±standard error.

It is worth noting that the CO_2 distribution pattern varied between bedroom and lounge space. The peak phase shifted from nighttime periods in bedroom to daytime periods in communal spaces. One can gain insight about the time diary of the elderly residents from CO_2 data. CO_2 is a major type of human metabolic emissions, which would in turn be associated with the number of occupants, duration of residency, and intensity of activity performed. Figure 6 indicated that the elderly residents spent most of their daytime periods in communal areas rather than bedrooms.

(2) Thermal comfort

As illustrated in Figure 8, the outdoor hourly mean dry bulb temperature ranged from 10.8 °C to 14.9 °C, with an average across the five weekdays of 12.5 °C. The temperature profile measured at bedroom was maintained between 20 °C and 23 °C. The temperatures of communal space were comparable with values recorded at the bedroom, but one can observe significant peaks during afternoon periods.

Compared to outdoor relative humidity (RH) profile, the indoor values fluctuated in a narrow range. The daily 1-h mean RH of the bedroom ranged from 44% to 51%, with an average across the five monitoring days of 48%. The values recorded at communal area ranged from 50% to 55%, with an average of 52%.

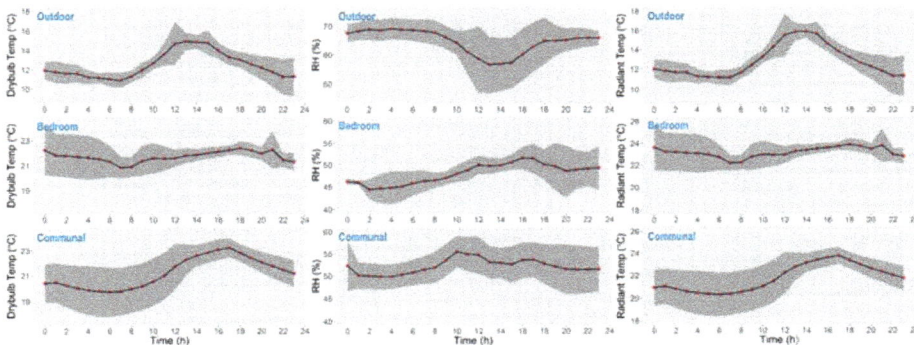

Figure 8. Daily 1-h mean (±standard error): dry bulb temperature (**left**); relative humidity (**middle**); and radiant temperature (**right**) measured at outdoor, bedroom and communal space over the full monitoring period. The shaded areas represent variability of results as ±standard error.

The Thermal Comfort Tool developed by the Center for the Built Environment (CBE) at the University of California, Berkeley, was applied to examine if Adare SRS provides a comfortable thermal environment for the elderly residents [29]. The PMV model was selected because: (i) the elderly have limited ability to control their thermal comfort; and (ii) the indoor air speed is negligible in experimental period because the house relies on air infiltration/exfiltration for ventilation purpose on winter days. As illustrated in Figure 7, the mean radiant temperatures followed the dry bulb temperature profiles through the monitoring period. The weekly average values, 23.1 °C and 21.8 °C for bedroom and communal space, respectively, were used to estimate the thermal condition. There is no discernible air flow in indoor environment. The low air flow can be attributable to the fact that windows and doors are usually kept closed in winter days and house relies on air infiltration/exfiltration for ventilation purpose. The elderly residents are usually in sedentary status (metabolic rate = 1 met) and wearing winter clothes (clothing level = 1 clo) during the monitoring period.

Although the thermal condition of both bedroom and communal space are in the lower range of thermal comfort zone defined by ANSI/ASHRAE Standard 55-2013 [30] (Figure 9), a warmer environment is recommended. This is because: (i) the standard comfort zone is the thermal condition preferred by healthy adults, rather than vulnerable elderly residents; and (ii) previous studies indicate that the 20–24 °C comfort zone is not warm enough for older adults and older adults generally prefer a warmer environment than younger subjects [31,32].

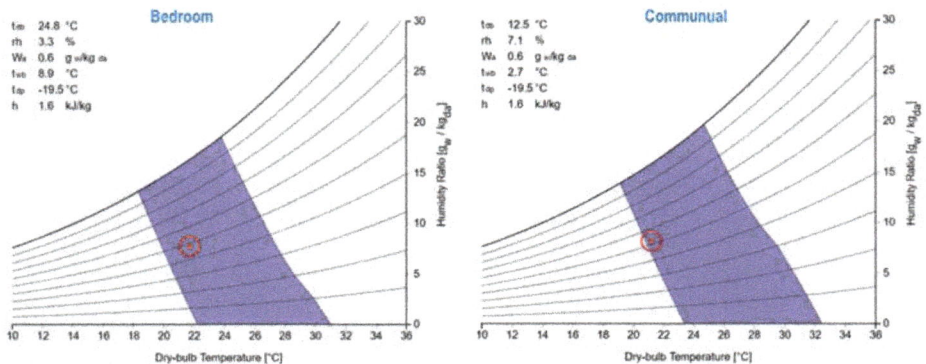

Figure 9. Thermal condition of: the bedroom (**left**); and the communal space (**right**) in psychometric chart. The shaded areas represent comfort zone boundary. The averaged values for the monitoring period were used in this estimation.

(3)   Concentrations of particulate matters

Previous studies suggest association of particulate matter (PM) and its compounds with health problems in the elderly such as the acute respiratory inflammation, pneumonia, asthma chronic obstructive pulmonary disease, autonomic cardiac dysfunction, renal and cognitive deficit, and cardiovascular and respiratory mortality [33–36]. For both $PM_{0.3–2.5}$ and $PM_{0.3–10.0}$, the indoor concentrations were lower than outdoor levels (Figure 10). This implies that the building envelope can protect the elderly residents against outdoor PM pollutants.   One can observe elevated $PM_{0.3–10.0}$ concentration measured at communal area during day times, which is consistent with CO_2 concentration profile.

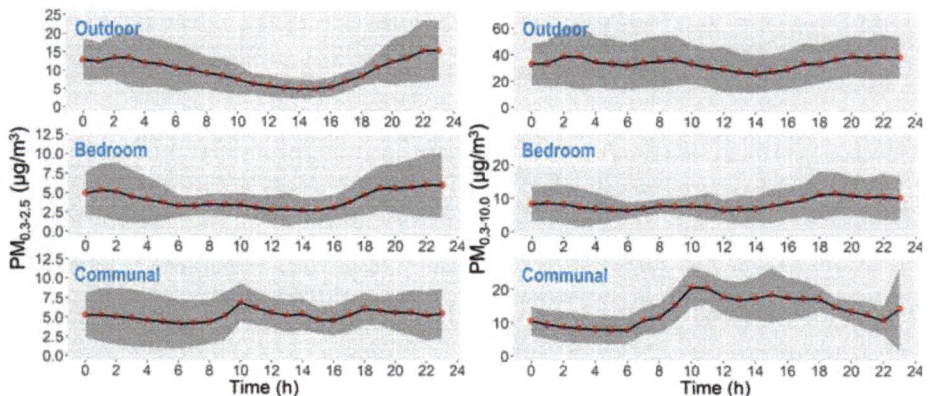

Figure 10. Daily 1-h mean (±standard error): $PM_{0.3–2.5}$ (**left**); and $PM_{0.3–10.0}$ (**right**) concentrations measured at outdoor, bedroom and communal space over the full monitoring period. The shaded areas represent variability of results as ±standard error.

Menders and his colleagues investigated a wide of range of indoor environmental quality parameters (including CO_2 and particle concentrations, thermal condition parameters) of elderly care centers in GERIE study (Geriatric study on health effects of air quality in nursing homes) [37–40]. The aged care residents were exposed to higher level of CO_2, but warmer temperature in this case study than GERIE project. The elevated CO_2 concentrations and temperature can be attributable to the reduced indoor-outdoor air exchange in winter season. The particle concentrations reported in previous studies are two to three times higher than the values we observed in this work. The low air exchange rate and better air quality outdoors probably contribute to the reduction of particle concentrations.

This section explored spatial and environmental conditions of a selected aged care facility. The temperature, relative humidity and air contaminants were monitored to examine if the minimum standard of indoor-environmental quality had been met. Nonetheless, the objective monitoring outcomes themselves do not necessarily reflect users' subjective comfort or environmental experiences. There is a clear need to address subjectivity. A question arises: How can the environmental data collected contribute to improving spatial settings of aged care facilities? Based on the need, the following section will analyse the subjectivity and propose the environmental experience design research framework.

4. Environmental Experience Design Research Framework

Due to the growth of global ageing population, increasing attentions have been drawn to accommodate the needs of the elderly's experience design [41]. Lawton initially raises the significance of the "reciprocal relationship between the elderly and their environment" [42,43]. Elf et al. also indicate the importance of the built environment that fulfils modern care vision embracing "autonomy, relatedness and competence" [44,45]. Nonetheless, Holahan and Moore et al. criticise the lack of theoretical research framework on person-environment nexuses in environmental psychology [46,47]. The gap may be filled by combining theories and practices around emotion, user experience design, experience design and environmental psychology—those that were briefed in the previous sections. In this study, the integrated research approach will be termed "environmental experience design" (EXD) which aims to provide a framework to identify design objectives, to analyse user perception and to propose design strategies. In the EXD research framework, both objective physical parameters (e.g., environmental quality data) and subjective user perception (e.g., human emotion) are brought into account for the development of spatial design strategies. As a case study, this paper intends to explore the proposed research framework through the application to a selected aged care facility located in Victoria. Based on the literature review of theories related to this study, the following sections unveil relevant design objectives towards the establishment of the EXD research framework.

4.1. Environmental Experience Design of Freedom

The current care model of the selected Adare SRS enables the residents to enjoy their freedom with minimum restrictions. Lawton and Nahemow summarise that origins of the elderly's stress are mainly due to the lack of "convenience to services, friends, and relatives, characteristics of structure and availability of social services" [42]. However, all these stated concerns may be solved with the notion of providing freedom. Parmelee and Lawton suggest that one of the aspects of human-environment relations in the elderly's late life depends on "the dialectic of autonomy" [48]. Autonomy is defined as "a state in which the person is, or feels, capable of pursuing life goals by the use of his or her own resources" which indicates "freedom of choice, action, and self-regulation of one's life space—in other words, the perception of and capacity for effective independent action" [48]. Nevertheless, "freedom of choice, action and self-regulation" are intertwined. As Dubos argues, "people want to experience the sensory, emotional, and spiritual satisfactions that can be obtained only from an intimate interplay, indeed from an identification with the places which [they] live" [49]. However, the mere consideration of the experience of freedom with accessibility to social infrastructure seems to be inadequate, architectural design also plays an important role of realising the true freedom. Open floor planning could be considered as a fundamental design approach to maximising the use of limited spaces while minimising partitions. Movable partitions may also enhance the space use efficiency and flexibility. Unnecessary level changes on the floors may reduce or eliminate physical and visual barriers associated with the elderly movements (e.g., with a wheelchair, wheeled walker, scooter and mobile hoist) and other daily life activities including social networking. Van Steenwinkel et al. advocate that this open spatial articulation enables more social interactions amongst residents, but also affords "spatial generosity" through combining "surface area, room to maneuver and variety of places" [50].

4.2. Environmental Experience Design of Connection to Natural Environment

The natural environment is a source of "interest, fascination and affection" and it may serve as a medium which directly or indirectly influences human psychology in the built environment [51]. Human relationship with nature may also be considered as the important part of environmental experience design. In such case, direct experience may involve the built environment connecting with self-sustaining natural elements including daylight, vegetations, animals and ecosystems, while indirect experience may refer to partial natural elements, such as potted plants and fountains, symbolic-represented natural features through images, photos and metaphors. Salingaros and Masden argue that "we instinctively crave physical and biological connection to the world. The human perceptual mechanisms through which these processes establish our relationship and response to both architecture and the built environment" [52]. Kellert uses the term "building for life" to highlight the significance of the re-establishment of the positive relationship between occupants and nature in the built environment [53]. Such positive relationship may be reinforced through various architectural design approaches, such as allowing daylight access, natural sound transmission and botanic fragrance perception for the sake of fulfilling the inherent need of humans to connect with nature [54].

4.3. Environmental Experience Design of Belongingness

Strong physical and psychological needs are retained to call places as homes regardless of propensity of mobility [55]. However, the feeling of "placelessness" may exist in the elderly's emotions [56,57]. Against the sense of placelessness, Relph emphasises the importance of maintaining the sense of "placeness" which can be "sources of safety and identity for individuals and for groups of people" [57]. Salingaros and Masden share the similar view of connecting to the built environment by having "a deep sense belonging to it" [52]. As far as architectural design approach is concerned, a welcoming environment may need to create and provide a sense of "placeness", which implies that the architectural space needs to be vibrant and flexible to respond to the elderly's changing needs, interests and abilities.

4.4. Environmental Experience Design of Individual Dignity

Designing environmental experience of individual dignity may enhance the elderly's sense of control in the built environment given and help them feel secured and comfortable physically and psychologically. Andrews and Philips argue that to provide a healthy living environment is vital to understand how the elderly establish their feelings of self-esteem in the aged-care facility [58]. Kitwood also points out a notion regarding "the uniqueness of each person, subjectivity, and relatedness", however, "treating human beings consistently as persons is so rare in everyday life" [59]. Architecturally, environmental experience design of individual dignity may be performed as non-hierarchical, manageable and controllable spaces.

The aforementioned physical and psychological parameters such as freedom, connection to natural environment, belongingness and individual dignity were considered as general environmental experience design objectives (Table 1). In the table, the user experiences and associated spatial design strategies and/or solutions were enumerated.

Table 1. The general environmental experience design objectives.

Selected EXD Objectives	Analysed User Experiences	Potential Spatial Design Strategies and/or Solutions
Freedom	Feeling in control of their own actions rather than being controlled by others	Open floor planning Movable partitions Level difference avoidance
Connection to natural environment	Feeling connected with nature in the built environment	Direct and indirect experiences of natural environment in relation to health and well-being
Belongingness	Feeling an intimate contact with the place they live in	Vibrant and flexible spaces to correspond with the elderly's changing needs, interests and abilities
Individual dignity	Feeling being respected by others	Non-hierarchical, manageable and controllable spaces

4.5. The Environmental Experience Design Research Framework

The aim of structuring this EXD framework is to meet physical and psychological needs and demands from user perspectives. To achieve it, the methodology of function analysis system is introduced. Such methodology is important to identify "the performance of a user function" and refine the design procedure to "fulfil a user requirement" [60]. It determines by asking what functions that users need and how designers satisfy these. Consequently, the Function Analysis System Technique (FAST) diagram is applied for taking project functions and arranging them in a logical order [60]. It aids in clarifying problems objectively through visualising and understanding logical relationships between related functions. Such FAST diagram may help to explore the elderly users' physical and psychological needs and demands. As an example, it was generated to explore functions that relate to the elderly users' physical and psychological activation for their health and wellbeing (Figure 11).

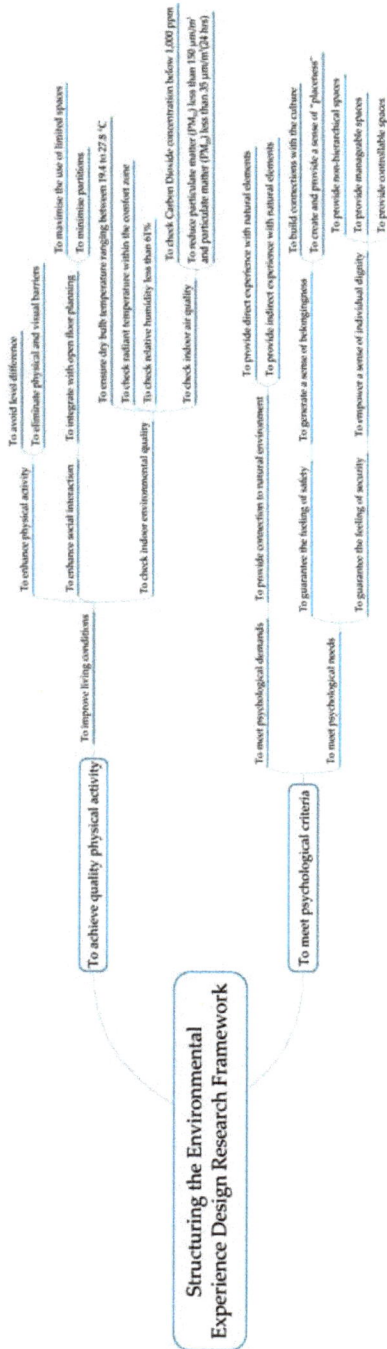

Figure 11. Environmental experience design function analysis system technique (FAST) diagram example.

The function analysis results may need to be converted into a medium that provides a design guideline to which stakeholders involved in design development can apply. Accordingly, in this study, an EXD research framework matrix was proposed, encompassing related design criteria, design settings, objectives, design elements, opportunities and design solutions. For the demonstration purpose, an EXD research framework matrix for the design improvement of aged care facilities was developed based on the previous function analysis results (Table 2). The EXD matrix serves as a diagnostic tool to recap the effective functions and transform them into design solutions that can be applied to upgrading new and existing buildings according to users' physical and psychological needs and demands.

Table 2. Environmental experience design diagnostic matrix.

Design Criteria	Design Settings	Objectives	Design Elements	Opportunities	Design Solutions
Physical Activation Criteria	Natural setting	Natural variability	Sounds	Providing pleasing effects	Euphonic sounds increase
			Sounds	Providing soothing effects	Chaotic sounds reduction
			Touch responses	Interaction with vegetation	Garden walking pathways
			Touch responses	Interaction with animals	Freedom of feeding pets
			Smell responses	Olfactory stimuli	Different scented plants
			Ageing and changing	The rhythm of life	Seasonal plants assembly
		Natural harmony	Complex order	Rich sensory information	Different types of flowers and trees gardening
			Patterned wholes	Intriguing balance between boring and overwhelming	Symmetric and fractal geometries of integrated pavements
		Naturalness	Outdoor gardens	Ecological connectivity	Communal spaces in the garden
			Garden crafts	Culture and ecology integration	The elderly engaged gardening
	Built environment setting	Spatial adaptability	Present spatial needs	Achieving the elderly's satisfactions	Calming colours of wall and ceiling paintings
					Natural textured material of floorings
					Two layered curtains: gauze and fabric curtains
					Freedom of bringing personal belongings
					Lockable interior doors
					Keep gardens open
		Spatial flexibility	Future spatial needs	Pliable temporal limitation	Moveable partitions
		Spatial durability	Ease of maintenance spaces	Cleanness guarantee	Flooring with vinyl
		Spatial arrangement	Spatial requirement	Friendly, comfortable and welcoming living conditions	Kitchen adheres to dining areas
					Communicative living rooms
					Bedrooms with private bathroom
					Multi-activity rooms
					Corridors with hand rails
					Clinic and therapy areas
					A nurse's station
			Spatial relationship		Kitchen next to dining room
					Nurses station away from dining room
					Nurses station away from reception and entrance
					Nurses station away from high-care residents' bedrooms
					Nurses station away from clinic and therapy areas
					Nurses station away from staff office
					Low-care elderly's bedrooms next to activity rooms
			Spatial layout		Using bubble diagrams
		Indoor environmental quality	Dry bulb temperature	Elderly wellbeing, comfort and health	Direct heat exposure avoidance: 19.4 to 27.8 °C
			Radiant temperature		Shading provision: within the comfort zone
			Relative humidity		Fresh air ventilation: less than 61%
			Indoor air quality		Fresh air ventilation: CO_2 concentration below 1,000 ppm
					Particulate matter (PM_{10}) less than 150 $\mu m/m^3$ (in 24 hrs)
					Particulate matter (PM_{10}) less than 35 $\mu m/m^3$ (in 24 hrs)
Psychological Activation Criteria	Psychological demands	Emotional wellness	Anxiety reduction	Harmful effects on health and the overall quality of life	Respect the elderly's personal choices and decisions
			Angry reduction		
			Depression reduction		
		Passive participation	Activating physical walking	Increase engagement and participation	Low-care elderly's bedrooms in a distance to living rooms
			Visual connection to activity rooms		Lay-outing the low-care elderly's bedrooms adjacent to activity rooms
	Psychological needs	Security	Empowering a sense of individual dignity	Being free from injury	Non-hierarchical spaces
					Manageable spaces
					Controllable spaces
		Safety	Generating a sense of belongingness	Being protected from risk	Locating residential care facilities in the neighbourhood
					Rooms have visual connections with outdoor
					Stimulating the elderly to bring vernacular objects
					Stimulating the elderly engaged gardening
					A space with the passage of time
		Active participation	Barrier-free spatial conditions	Physical activity stimulation	Level difference avoidance
					Slip-resistant and firm flooring surfaces
					Physical and visual barriers elimination
					Wide interior doors, corridors and turning spaces associated with the elderly movements
			Open floor planning	Social interaction	Maximising the use of limited spaces
					Minimising partitions
			Multi-activity rooms' design	Restorative activity	Evoking and developing the elderly's interests

5. Conclusions

This study led to the creation of a new terminology "environmental experience design" (EXD) and the research framework through literature reviews of theories on emotion, user experience design, experience design and environmental psychology, as well as studies on existing spatial settings and indoor environmental quality of an aged care facility selected in Victoria, Australia. Through the case study of a selected aged care facility (i.e., Adare SRS), various physical environmental data (including dry bulb temperature, radiant temperature, relative humidity and indoor air quality) have been collected and analysed. However, the objective outcomes from monitoring itself do not necessarily reflect users' subjective comfort or environmental experiences; therefore, human subjective environmental experiences should be examined further. Accordingly, the EXD approach was proposed considered an attempt that affiliates experience design and environmental psychology with the creation of the built environment that requires to accommodate user needs and demands. The EXD explores design essentials beyond physical data-driven interpretations. In this paper, the above-mentioned EXD objectives provided a base for further execution of related function analysis and matrix development. Using the Adare SRS case, the Function Assessment System Technique (FAST) diagram and the EXD diagnostic matrix for the elderly users' physical and psychological activation were demonstrated. The EXD matrix served as a diagnostic tool to recap the effective functions and transform them into design solutions. It can be applied to upgrading new and existing buildings according to users' physical and psychological needs and demands.

The quality of space needs to be evaluated by users themselves. In this paper, the EXD research framework matrix was developed with due consideration of the elderly users' behavioural patterns observed; however, their direct opinions over spaces were not collected. Therefore, the questionnaires about their post-EXD solutions and potential outcomes should be gathered and analysed. The EXD approach can also be applied to any other spatial design decisions under various environmental conditions. Therefore, the application to different architectural topologies and environmental conditions should be conducted.

Acknowledgments: The authors would like to thank the Adare SRS for providing access to the facility for research activities and the Thrive Research Hub for providing measuring devices for collecting physical environmental data.

Author Contributions: The first author led overall research activities and contributed to structuring the proposed environmental experience design diagnostic framework. The second author conducted spatial data collation of an aged care facility selected. The third author dealt with the indoor environmental quality assessment. The fourth author contributed to the total editorial coordination and arrangement.

Conflicts of Interest: The authors declare no conflict of interest.

References and Notes

1. Crozier, R. *Manufactured Pleasures: Psychological Responses to Design*; Manchester University Press: New York, NY, USA, 1994.
2. Dewey, J. *Experience and Education*; Collier: New York, NY, USA, 1963.
3. McCarthy, J.; Wright, P. *Technology as Experience*; MIT Press: Cambridge, MA, USA, 2004.
4. Hassenzahl, M. *Experience Design: Technology for All the Right Reasons*; Morgan & Claypool Publishers: San Rafael, CA, USA, 2010.
5. Norman, D.A. *The Psychology of Everyday Things*; Basic Books: New York, NY, USA, 1988.
6. O'Sullivan, E.L.; Spangler, K.J. *Experience Marketing: Strategies for the New Millennium*; Venture Pub: Jersey Shore, PA, USA, 1998.
7. Pine, B.J.; Gilmore, J.H. *The Experience Economy: Work Is Theatre & Every Business a Stage*; Harvard Business School Press: Boston, MA, USA, 1999.
8. Berridge, G. *Events Design and Experience*; Butterworth-Heinemann: Burlington, NJ, USA, 2007.
9. Kellert, S.R.; Heerwagen, J.; Mador, M. Preface. In *Biophilic Design: The Theory, Science and Practice of Bringing Buildings to Life*; Kellert, S.R., Heerwagen, J., Mador, M., Eds.; John Wiley & Sons: New York, NY, USA, 2007; pp. vii–ix.
10. McLellan, H. Experience design. *Cyber Psychol. Behav.* **2000**, *3*, 59–69. [CrossRef]
11. Salingaros, N.A. *A Theory of Architecture*; Umbau-Verlag: Solingen, Germany, 2006.
12. Frumkin, H. Beyond toxicity: Human health and the natural environment. *Am. J. Prev. Med.* **2001**, *20*, 234–240. [CrossRef]
13. Gifford, R. *Environmental Psychology: Principles and Practice*; Allyn and Bacon: Boston, MA, USA, 1997.
14. Nagar, D. *Environmental Psychology*; Concept Publishing Company: New Delhi, India, 2006.
15. Verderber, S.; Fine, D.J. *Healthcare Architecture: In an Era of Radical Transformation*; Yale University Press: New Haven, CT, USA, 2000.
16. Department of Housing and Construction. *Guidelines for the Provision of Nursing Home Facilities*; Australian Government Publishing Service: Canberra, Australia, 1983.

17. Chau, H.W. De-Institutionalisation of Aged Care Residential Facilities in Australia: Aldersgate Village in Adelaide and Wintringham Facilities in Melbourne as Case Studies. In Proceedings of the Society of Architectural Historians Australia and New Zealand, Sydney, Australia, 7–10 July 2015; Hogben, P., O'Callaghan, J., Eds.; SAHANZ: Sydney, Australia, 2015; Volume 32, pp. 82–91.

18. Carboni, J. Homelessness among the elderly. *J. Gerontol. Nurs.* **1990**, *16*, 32–37. [CrossRef] [PubMed]

19. Maslow, A. A Theory of human motivation. *Psychol. Rev.* **1943**, *50*, 370–396. [CrossRef]

20. Springer, D.; Brubaker, T.H. *Family Caregivers and Dependent Elderly: Minimizing Stress and Maximizing Independence*; Sage Publications: Beverly Hills, CA, USA, 1984.

21. Kidd, B.J.; Kidd, L.A. *The Image of Home: An Alternative to Institutional Design for Elderly People in Nursing Homes*; The Centre for Applied Research on the Future, The University of Melbourne: Melbourne, Australia, 1987.

22. Kidd, B.J. *Hostel Design Guidelines*; Australian Government Publishing Service: Canberra, Australia, 1988.

23. Victorian Government Department of Human Services. *Aged Care Residential Services Generic Brief*; Victorian Government Department of Human Services: Melbourne, Australia, 2000.

24. Murphy, S. *Knight Frank Health and Aged Care*, lecture notes; The University of Melbourne; delivered 12 May 2017.

25. State of Victoria. Supported Residential Services. Available online: https://www2.health.vic.gov.au/ageing-and-aged-care/supported-residential-services (accessed on 12 July 2017).

26. Allen, J.G.; Naughton, P.M.; Satish, U.; Santanam, S.; Vallarino, J.; Spengler, J.D. Associations of cognitive function scores with carbon dioxide, ventilation, and volatile organic compound exposures in office workers: A controlled exposure study of green and conventional office environments. *Environ. Health Perspect.* **2016**, *124*, 805–812. [CrossRef] [PubMed]

27. Satish, U.; Mendell, M.J.; Shekhar, K.; Hotchi, T.; Sullivan, D.; Streufert, S.; Fisk, W.J. Is CO_2 an indoor pollutant? Direct effects of low-to-moderate CO_2 concentrations on human decision-making performance. *Environ. Health Perspect.* **2012**, *120*, 1671–1677. [CrossRef] [PubMed]

28. American Society of Heating, Refrigerating and Air-Conditioning Engineers. *Standard 62.1 User's Manual: Based on ANSI/ASHRAE Standard 62.1-2016, Ventilation for Acceptable Indoor Air Quality*; ASHRAE: Atlanta, GA, USA, 2016.

29. Hoyt, T.; Stefano, S.; Alberto, P.; Dustin, M.; Kyle, S. CBE Thermal Comfort Tool. 2013. Available online: http://cbe.berkeley.edu/comforttool/ (accessed on 28 August 2017).

30. American Society of Heating, Refrigerating and Air-Conditioning Engineers. *ANSI/ASHRAE Standard 55-2013: Thermal Environmental Conditions for Human Occupancy*; ASHRAE: Atlanta, GA, USA, 2013.

31. Kimura, M.; Ikaga, T.; Hirayama, Y. Effects of radiant cooling and heating system on thermal comfort and productivity in elderly care facilities. In Proceedings of the 12th International Conference on Indoor Air Quality and Climate, Austin, TX, USA, 5–10 June 2011; Curran Associates, Inc.: New York, NY, USA, 2011.

32. Schellen, L.; van Marken Lichtenbelt, W.D.; Loomans, M.G.; Toftum, J.; de Wit, M.H. Differences between young adults and elderly in thermal comfort, productivity, and thermal physiology in response to a moderate temperature drift and a steady-state condition. *Indoor Air* **2010**, *20*, 273–283. [CrossRef] [PubMed]

33. Mehta, A.J.; Zanobetti, A.; Bind, M.A.; Kloog, I.; Koutrakis, P.; Sparrow, D.; Vokonas, P.S.; Schwartz, J.D. Long-term exposure to ambient fine particulate matter and renal function in older men: the veterans administration normative aging study. *Environ. Health Perspect.* **2016**, *124*, 1353–1360. [CrossRef] [PubMed]

34. Weuve, J.; Puett, R.C.; Schwartz, J.; Yanosky, J.D.; Laden, F.; Grodstein, F. Exposure to particulate air pollution and cognitive decline in older women. *Arch. Intern. Med.* **2012**, *172*, 219–227. [CrossRef] [PubMed]

35. Han, Y.; Zhu, T.; Guan, T.; Zhu, Y.; Liu, J.; Ji, Y.; Gao, S.; Wang, F.; Lu, H.; Huang, W. Association between size-segregated particles in ambient air and acute respiratory inflammation. *Sci. Total Environ.* **2016**, *565*, 412–419. [CrossRef] [PubMed]

36. Power, M.C.; Elbaz, A.; Beevers, S.; Singh-Manoux, A. Traffic-related air pollution and cognitive function in a cohort of older men. *Environ. Health Perspect.* **2011**, *119*, 682–687. [CrossRef] [PubMed]

37. Aguiar, L.; Mendes, A.; Pereira, C.; Neves, P.; Mendes, D.; Teixeira, J.P. Biological air contamination in elderly care centers: Geria project. *J. Toxicol. Environ. Health Part A Curr. Issues* **2014**, *77*, 944–958. [CrossRef] [PubMed]

38. Almeida-Silva, M.; Wolterbeek, H.T.; Almeida, S.M. Elderly exposure to indoor air pollutants. *Atmos. Environ.* **2014**, *85*, 54–63. [CrossRef]

39. Bentayeb, M.; Norback, D.; Bednarek, M.; Bernard, A.; Cai, G.; Cerrai, S.; Eleftheriou, K.K.; Gratziou, C.; Holst, G.J.; Lavaud, F.; et al. Indoor air quality, ventilation and respiratory health in elderly residents living in nursing homes in Europe. *Eur. Respir. J.* **2015**, *45*, 1228–1238. [CrossRef] [PubMed]

40. Mendes, A.; Pereira, C.; Mendes, D.; Aguiar, L.; Neves, P.; Silva, S.; Batterman, S.; Teixeira, J.P. Indoor air quality and thermal comfort—Results of a pilot study in elderly care centers in Portugal. *J. Toxicol. Environ. Health Part A Curr. Issues* **2013**, *76*, 333–344. [CrossRef] [PubMed]

41. Department of Economic and Social Affairs. Available online: http://www.un.org/en/development/desa/population/publications/pdf/ageing/WPA2015_Report.pdf (accessed on 18 August 2017).

42. Lawton, M.; Nahemow, L. Ecology and the aging process. In *The Psychology of Adult Development and Aging*; Eisdorfer, C., Lawton, M., Eds.; American Psychological Association: Washington, DC, USA, 1973; pp. 619–674.

43. Lawton, M.P. *Environment and Aging*, 2nd ed.; Albany: New York, NY, USA, 1986.

44. Deci, E.L.; Ryan, R.M. The "what" and "why" of goal pursuits: human needs and the self-determination of behavior. *Psychol. Inq.* **2000**, *11*, 227–268. [CrossRef]

45. Elf, M.; Fröst, P.; Lindahl, G.; Wijk, H. Shared decision making in designing new healthcare environments—Time to begin improving quality. *BMC Health Serv. Res.* **2015**, *15*, 114–120. [CrossRef] [PubMed]

46. Holahan, C. Environmental psychology. *Annu. Rev. Psychol.* **1986**, *37*, 381–407. [CrossRef]

47. Moore, K.; VanHaitsma, K.; Curyto, K.; Saperstein, A. A pragmatic environmental psychology: A metatheoretical inquiry into the work of M. Powell Lawton. *J. Environ. Psychol.* **2003**, *23*, 471–482. [CrossRef]

48. Parmelee, P.; Lawton, M.P. The design of special environments for the aged. In *Handbook of the Psychology of Aging*, 3rd ed.; Birren, J.E., Ed.; Academic Press: London, UK, 1990; pp. 464–488.

49. Dubos, R. *The Wooing of Earth*; Scribner: New York, NY, USA, 1980.

50. Van Steenwinkel, I.; Dierckx de Casterlé, B.; Heylighen, A. How architectural design affords experiences of freedom in residential care for older people. *J. Aging Stud.* **2017**, *41*, 84–92. [CrossRef] [PubMed]

51. Kaplan, R.; Kaplan, S. *The Experience of Nature: A Psychological Perspective*; Cambridge University Press: Cambridge, UK, 1989.

52. Salingaros, N.; Masden, K. Neuroscience the natural environment and building design. In *Biophilic Design: The Theory, Science and Practice of Bringing Buildings to Life*; Kellert, S., Heerwagen, J., Mador, M., Eds.; John Wiley: New York, NY, USA, 2008; pp. 59–83.

53. Kellert, S. *Building for Life: Designing and Understanding the Human-Nature Connection*; Island Press: Washington, DC, USA, 2005.

54. Ryan, C.O.; Browning, W.D.; Clancy, J.O. Biophilic design patterns. *Archinet IJAR* **2014**, *8*, 62–75. [CrossRef]

55. Kellert, S. Dimensions, element, and attributes of biophilic design. In *Biophilic Design: The Theory, Science and Practice of Bringing Buildings to Life*; Kellert, S., Heerwagen, J., Mador, M., Eds.; John Wiley: New York, NY, USA, 2008; pp. 3–19.

56. Yu, J.; Rosenberg, M. "No place like home": Aging in post-reform Beijing. *Health Place* **2017**, *46*, 192–200. [CrossRef] [PubMed]

57. Relph, E. *Place and Placelessness*; Pion: London, UK, 1976.

58. Andrews, G.J.; Phillips, D.R. *Ageing and Place: Perspectives, Policy, Practice*; Routledge: New York, NY, USA, 2005.

59. Kitwood, T. *Dementia Reconsidered: The Person Comes First*; Open University Press: London, UK, 1997.

60. Dell'Isola, A.J. *Value Engineering: Practical Applications for Design, Construction, Maintenance & Operations*; Wiley: Kingston, UK, 1997.

MDPI

St. Alban-Anlage 66

4052 Basel, Switzerland

Tel. +41 61 683 77 34

Fax +41 61 302 89 18

http://www.mdpi.com

Sustainability Editorial Office

E-mail: sustainability@mdpi.com

http://www.mdpi.com/journal/sustainability

www.ingramcontent.com/pod-product-compliance
Lightning Source LLC
Chambersburg PA
CBHW051843210326
41597CB00033B/5763